THE CH/π INTERACTION

Methods in Stereochemical Analysis

Series Editor

Alan P. Marchand, Denton, Texas, USA

Advisory Board

A. Greenberg, Charlotte, North Carolina, USA
I. Hargittai, Budapest, Hungary
Chen C. Ku, Shanghai, China
J. Liebman, Baltimore, Maryland, USA
E. Lippmaa, Tallinn, Estonia
S. Sternhell, Sydney, Australia
Y. Takeuchi, Tokyo, Japan
F. Wehrli, Philadelphia, Pennsylvania, USA
D. H. Williams, Cambridge, UK
N. S. Zefirov, Moscow, Russia

THE CH/π INTERACTION
Evidence, Nature, and Consequences

MOTOHIRO NISHIO
MINORU HIROTA
YOJI UMEZAWA

WILEY-VCH

New York / Chichester / Weinheim / Brisbane / Singapore / Toronto

Copyright © 1998 by Wiley-VCH, Inc. All rights reserved.

Published simultaneously in Canada.

Library of Congress Cataloging-in-Publication Data:

Nishio, Motohiro, 1936– .
    The CH/π interaction: evidence, nature, and consequences/
Motohiro Nishio, Minoru Hirota, Yoji Umezawa.
        p.   cm.—(Methods in stereochemical analysis)
    Includes bibliographical references and index.
    ISBN 0–471–25290–5 (cloth: alk. paper)
    1. Chemistry, Physical organic. 2. Hydrogen bonding. I. Hirota, Minoru,
1933– . II. Umezawa, Yoji, 1950– . III. Title. IV. Series: Methods in
stereochemical analysis (Unnumbered) QD476.N58   1998
541.2′24—dc21   97-41114

Printed in the United States of America.

10 9 8 7 6 5 4 3 2 1

PREFACE

Bond energies of normal covalent bonds in a molecule are 50–110 kcal mol^{-1}. Weaker intramolecular forces, such as the Coulombic force, the hydrogen bond, and the van der Waals force determine shape and behavior of molecules. Among the latter, the hydrogen bond is certainly one of the most important. Evidence has gradually accumulated that forces weaker than the ordinary hydrogen bond, the CH/O, CH/N, OH/π, and NH/π interactions, are also important. Among these, the CH/π interactions which has recently gained attention in the consideration of a variety of molecular phenomena, is dealt with in this treatise. The "CH/π bond" is the weakest among the hydrogen bonds, but has been found in a variety of substances to play important roles in their physical, chemical, and biological properties. The term "CH/π interaction" has been accepted because of its usefulness in describing the interaction between CH groups and π-systems, which is considerably stronger than expected from a mere dispersion mechanism.

In a variety of chemical and biochemical phenomena, we often encounter CH/π proximate structures or arrangements in molecules. Attractive interactions operating between CH and π in such cases could be understood as a combined effect of known interaction forces, (Coulombic, charge transfer, dispersion, so-called "hydrophobic," etc.), and may most appropriately be termed as CH/π interaction. In

Chapter 3 we define the term CH/π interaction in more detail. We hope that this book will clarify the use of this term and develop a common understanding of the CH/π interaction.

The possible role of the CH/π interaction in the conformation of molecules was first inspired by discussions with Professor Naoya Nakagawa (Tokyo University of Electrocommunications). We greatly appreciate his foresightedness. We are deeply indebted to Professor Eiji Osawa (Toyohashi University of Technology), one of the pioneers of this field, for encouraging us to write this book and for his advice and information during the writing. Sincere thanks are also due to all members of the authors' research groups for their skill and enthusiasm in carrying out the work described in this book.

We thank Dr. Kazuaki Harata (Institute of Biosciences and Biotechnology) and Professors Rocco Ungaro (Università di Parma), Katsuyuki Ogura (Chiba University), and Kazunori Odashima (University of Tokyo) for their crystallographic data and helpful discussions. Discussions with the following professors were also useful: the late Günter Snatzke (Rhur-Universität Bochum); Sir Derek H. R. Barton (Texas A & M University); J. Fraser Stoddart (University of Birmingham); Andrée Marquet (Université Paris 6); Hugh Felkin (Institut de la Chimie de Substances Naturelles, CNRS); Christian Roussel (Université Aix-Marseilles III); Andrew Streitwieser, Jr. and the late William G. Dauben (UCB); Paul R. von Schleyer (University of Georgia); Michinori Oki (Okayama Science University); Hiizu Iwamura, Hisashi Okawa, and Yasuhiro Aoyama (Kyushu University); Takayuki Shioiri (Nagoya City University); Shinichi Ueji (Kobe University); Yoichi Iitaka, Shigeo Iwasaki, Yoshinori Satow, and Keiji Kobayashi (University of Tokyo); Yutaka Fukuda (Ochanomizu University); Mikio Nakamura (Toho University); Tadashi Endo (Aoyama Gakuin University); Yumihiko Yano (Gunma University); Kenji Fujimori (Tsukuba University); the late Norio Kunieda (Osaka City University); Drs. Jun Uzawa, Kaoru Tsuboyama, and Sei Tsuboyama (Institute of Physical and Chemical Research); Sinichi Kondo (Institute of Microbial Chemistry); Ken Nishihata, Yoshio Kodama, Shoji Zushi, and Yasuo Takeuchi (Meiji Seika).

Information from the following was invaluable: Professors Helmut Sigel (Universität Basel), Max F. Perutz (MRC), Hideki Masuda (Nagoya Institute of Technology), Hiroyasu Imai (Hokuriku University), Toshimasa Ishida (Osaka College of Pharmacy), Hiroshi Shimizu (Gifu College of Pharmacy), Seiki Saito (Okayama University),

Kazuaki Yamanari and Tatsuya Kawamoto (Osaka University), Shohei Inoue (Tokyo Science University), and Akira Mori (Kyushu University) and Drs. Akira Tohara (Teikyo University), Yoshinobu Nagawa (Institute of Biosciences and Biotechnology), Kazumasa Honda (Institute of Materials and Chemical Research), Hiroshi Kimura (Mitsubishi Electric), Yoshihisa Inoue (Green Cross), and Naoyuki Amaya (Nihon Oil). Help from Professor Alan P. Marchand (University of North Texas) and Ms. Janet Teague-Nishimura in the editing of the manuscript, and assistance from Professors Kazuhisa Sakakibara and Hiroko Suezawa (Yokohama National University), and Kazue Yasufuku and Setsuno Igarashi (Meiji Seika) in the preparation of the manuscript was gratefully appreciated. We thank Drs. Yuzuru Akamatsu (Meiji Seika) and Tomio Takeuchi (Institute of Microbial Chemistry) for encouragement. M. N. in particular wishes to express his deep gratitude to Emeritus Professors Shigeru Oae (Tsukuba University), Yuzo Inoue (Kyoto University), Teiichiro Ito (Nihon University), and Koshiro Umemura (Toho University), and to Dr. Tomoko Shomura (Meiji Seika) and Kazuhiko Otomo (Nankodo). Without their kind help and encouragement, it is certain that this book could not have been completed.

MOTOHIRO NISHIO
MINORU HIROTA
YOJI UMEZAWA

CONTENTS

CHAPTER 1

INTRODUCTION

1.1. IMPORTANCE OF WEAK MOLECULAR INTERACTIONS

The properties of organic compounds have been described as the consequence of various kinds of chemical interactions—attractive or repulsive, strong or weak. Strong covalent bonds bind the atoms together in a molecule, whereas weak noncovalent interactions are important in deciding the shape, or conformation, of the molecule.

Noncovalent forces also play an important role in chemical reactions and molecular recognitions, and in regulating biochemical processes. Specificity and efficiency in these chemical processes are achieved by intricate combinations of weak intermolecular interactions of various sorts.

Dispersion, or van der Waals force, is the weakest among the many types of secondary interaction forces. It arises from fluctuating charges caused by the nearness of molecules, independent of their polarity, and produces nonspecific attraction. The strength of the dispersion force depends sharply upon the distance between the interacting groups, and it is inversely proportional to the sixth power of distance ($1/r^6$).

1.1.1. Hydrogen Bonds

Among intermolecular interactions, the hydrogen bond[1] is one of the most abundant, especially in the dynamic processes involved in biochemical reactions. The enthalpy of the hydrogen bond between proton donor groups such as OH or NH (hard acids) and electronegative atoms such as O or N (hard bases) is within the range of $3-7 \, kcal \, mol^{-1}$.[2] In the past few decades, evidence has accumulated to show that other attractive forces ($2-4 \, kcal \, mol^{-1}$), weaker than the ordinary hydrogen bond, are also ubiquitous, including XH/π ($X = O$ or N: hard acids versus soft bases)[3] and CH/n (n = lone pair electrons: soft acids versus hard bases) interactions.[4]

In recent years, it has gradually become accepted that a still weaker attractive force, the CH/π interaction,[5] exists in a variety of chemical and biochemical phenomena. We may regard the CH/π interaction as the weakest hydrogen bond occurring between a soft acid (CH) and a soft base (π-electrons). Energy from the hydrogen-bond-type interactions decrease approximately in the order hydrogen bond $> XH/\pi \approx CH/n > CH/\pi$. The CH/π interaction has been suggested only recently.[6,7] The enthalpy of a one-pair CH/π interaction is presumed to be less than $1 \, kcal \, mol^{-1}$. A unique feature of this kind of attractive force is that a number of CH groups may participate simultaneously in the interaction with a π-base. Total energy of the interaction may be increased by simultaneously organizing CHs and/or π-groups into favorable structures. This point is crucial in understanding the role of weak secondary forces. Another characteristic, which is important when considering biochemical processes, is that the CH/π interaction can play its role in aqueous media as well as in nonpolar media.

1.1.2. Examples of the CH/π Interaction

The following example illustrates how the CH/π interaction works in a dynamically interacting system.

If a pair of different classes of compounds recognize each other, they sometimes assemble spontaneously. Thus, Stoddart et al. reported a synthesis of catenanes[8] from dimethylated β-cyclodextrin (DM-β-CD) and polyethers **1**. Table 1.1 lists the stability constants K_a and free energies of complexation $-\Delta G°$ for the intermediate $1:1$ complexes (pseudorotaxanes)[9] formed between DM-β-CD and a series of substrates **1** designed for the catenation.[10]

1

Table 1.1 indicates that complexation occurs more strongly as the effective surface of the aromatic ring of the ligand increases. Upfield proton NMR chemical shift changes are shown for H3 and H5 of the glucose moiety of DM-β-CD on complexation, indicating that the aromatic part of the guest is close to the inner walls of the cyclodextrin; this was supported by NOE experiments. A catenane product **2** was obtained by cyclizing **1** (Ar = 4,4'-dihydroxybiphenyl) and DM-β-CD (Fig. 1.1). Its crystal structure revealed that the aromatic unit of the synthetic macrocycle was included in the CD cavity and close to the CH groups inside the wall.

TABLE 1.1. Association Constants K_a and Free Energies of Complexation $-\Delta G°$ of 1 (R = H) with DM-β-CD (in D$_2$O, 25°C)

Ar	K_a (mol^{-1})	$-\Delta G°$ (kcal mol^{-1})	CIS^a
1,4-Dihydroxyphenyl (**A**)	350	3.50	-0.17
1,5-Dihydroxynaphthyl (**B**)	1180	4.20	-0.16
2,6-Dihydroxynaphthyl (**C**)	2800	4.70	-0.17
4,4'-Dihydroxybiphenyl (**D**)	37300	6.25	-0.20

^{a} Complexation-induced shift ($\Delta\delta$/ppm) for H3 of DM-β-CD.

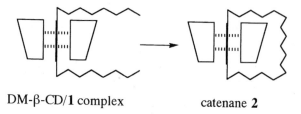

DM-β-CD/**1** complex   catenane **2**

Figure 1.1. CH/π interactions (dotted lines) in DM-β-CD/**1** complex and catenane **2**.

1.2. HISTORY

1.2.1. Activated C–H Groups Interact with π-Bases

It was first recognized that a CH group interacts favorably with a π-base when Tamres, in 1952, showed that aromatic compounds behaved as electron donors in the interaction with chloroform.[11] Thus, benzene and its analogs dissolve in chloroform exothermically. An aromatic compound which gave a higher heat of mixing with chloroform induced a larger shift in the infrared (IR) frequency of the O–D stretching band of the deuterated methanol. In 1956, Huggins and Pimentel found, by measurement of IR spectra, that the interaction of chloroform with benzene showed behavior consistent with the criteria for hydrogen bond systems.[12]

In 1957, Reeves and Schneider[13] showed, on the grounds of nuclear magnetic resonance (NMR) experiments, that the interaction of chloroform with benzene and olefins was a type of hydrogen bond. Nakagawa and Fujiwara[14] observed a high-field shift in the proton resonance of an ethinyl hydrogen of phenylacetylene by dilution with benzene; they attributed the phenomenon to a complex formation of the ethinyl CH with the aromatic π-system. Richards and Hatton[15] found that the ethinyl proton of propargyl chloride and phenylacetylene showed high-field shifts on dilution with aromatic solvents such as benzene or toluene.[16]

1.2.2. An Activated Methyl Group Interacts with π-Bases

In 1960, the interaction of acetone with benzene was studied by Schaefer, Schneider, and Buckingham by means of NMR measurements.[17] Nakagawa and Fujiwara,[18] and Shimizu[19] suggested, on the grounds of NMR, that the methyl groups in acetic acid, acetonitrile, nitromethane, and toluene derivatives interact with the solvent benzene. Yoshida and Osawa pointed out the possibility, by refractometric analyses, that ethyl acetoacetate and diethyl malonate form complexes with aromatic compounds.[20] However, this type of interaction was not considered to be typical and seems to be possible only when a CH_3 or CH_2 is next to an electron-withdrawing group such as halogens, carbonyl, or a phenyl group.

1.2.3. Ordinary alkyl Groups May Interact with π-Systems

In 1960, Pimentel and McClellan stated that " ... there surely remains some tendency for H bonding in methyl chloroform, and it might be observed in kinetic processes because of the extreme sensitivity of the reaction rate to the activation energy. This line of argument could be extended to C–H bonds in alkanes, aromatics, olefins ... ."

In the late 1970s, the existence of an attractive interaction between alkyl groups and π-bases was suggested, on the grounds of a conformational study on a series of compounds (R-X-Y-π).[7,21] In many cases the alkyl group (R in Fig. 1.2) has been shown to position itself close to the π-group (e.g., C_6H_5). Comparison of the experimental results with those obtained from molecular mechanics calculations[22] led the authors to propose that a weak but attractive force, the CH/π interaction, was working between an alkyl group and a π-system.

The suggestion found theoretical support in MO calculations for several supramolecular systems.[23] Calculations showed CH $\cdots \pi$ linear geometry to be the most stable (Fig. 1.3).

In such arrangements, one of the C-H bonds orients itself above an sp^2 carbon to give maximum overlap between the relevant orbitals. For the methane/ethylene supramolecular system, the contribution from the charge transfer ($\pi \rightarrow \sigma^*$) has been calculated to be

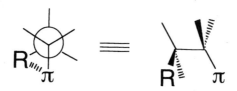

Figure 1.2. Preferred conformation of compounds, R-X-Y-π.

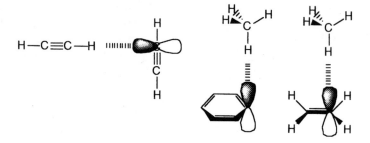

Figure 1.3. Preferred geometries for CH/π interaction.

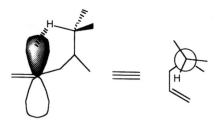

Figure 1.4. Stereochemical arrangement of the groups (CH vs. π) allowing enhancement of the CD intensity.

most important, the dispersion interaction being the second most important.[24]

Support for the contribution of charge-transfer interaction has come from considerations of the optical rotatory strength of unsaturated compounds (Fig. 1.4).[25] Thus, an appreciable enhancement in CD intensity has been shown to occur only when the geometrical arrangements of the groups allowing CH/π interaction is stereochemically possible.[26]

In 1980, Okawa et al. pointed out the possible importance of CH/π interaction to explain the stereoselectivity found in several coordination compounds. Preferential formation of one of the diastereoisomers has been observed for a series of metal complexes of 3-*l*-menthyloxy-1-phenyl-1,3-propanediones (Fig. 1.5).[27]

They interpreted the results in the context of the CH/π interaction. Thus, in the transition state, interaction of the *l*-menthyl group with the phenyl group is favored to give rise to selective formation of the cis-Δ product. The stereoselectivity became greater when the phenyl

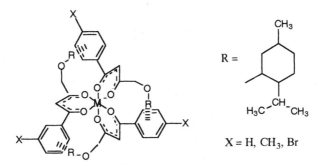

$R =$

$X = H, CH_3, Br$

Figure 1.5. Stereoselective formation of diastereomeric complexes.

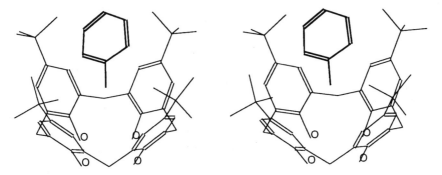

Figure 1.6. Toluene complex of a p-t-butylcalix[4]arene (stereo view).[30]

group was replaced by a naphthalene moiety and when the π-density of the aromatic ring increased by introduction of an electron-donating group to the aromatic group.

A suggestion for a possible role of the CH/π interaction in supramolecular chemistry was made in 1984, by Ungaro et al., in an X-ray crystallographic study of a toluene complex of p-t-butylcalix[4] arene.[28] It was reported that the methyl group of toluene was shown to point toward the cavity of the macrocycle, which is lined with many π-electrons, while the t-butyl groups of the host sandwiched the toluene aromatic ring (Fig. 1.6).[29]

The findings were followed by force-field calculations for a pyridine complex of a p-t-butylcalix[4]arene derivative,[31] supporting the CH/π interaction hypothesis. Several papers have since appeared on a variety of molecular species with presumed CH/π-interacted structures.[5, 32]

1.3. SCOPE OF THE MONOGRAPH

In the following chapters we discuss experimental results and consider the theoretical backgrounds which explain the possible relevance of the CH/π interaction. The nature and characteristics of the CH/π interaction is discussed in comparison with those of other types of weak interactions of a similar nature. As for the consequences, possible implications in conformations, chiroptical phenomena, and molecular recognition are emphasized. The molecular recognition includes selectivity in chemical reactions, supramolecular chemistry, and specificities in protein/ligand complexes. Chapter 10 is devoted to coordination chemistry and Chapter 12 presents future prospects.

REFERENCES

1. G. C. Pimentel and A. L. McClellan, *The Hydrogen Bond*, Freeman, San Francisco, 1960.
2. Reference 1, p. 224.
3. M. Oki, H. Iwamura, T. Onoda, and M. Iwamura, *Tetrahedron*, **24**, 1905 (1968); M. F. Perutz, *Phil. Trans. R. Soc.* A, **345**, 105 (1993).
4. A. Allerhand and P. von R. Schleyer, *J. Am. Chem. Soc.*, **85**, 1715 (1963); K. Morokuma, *Acc. Chem. Res.*, **10**, 294 (1977); R. Taylor and O. Kennard, *J. Am. Chem. Soc.*, **104**, 5063 (1982); T. Steiner and W. Saenger, *ibid.*, **114**, 10146 (1992); G. R. Desiraju, *Acc. Chem. Res.*, **24**, 290 (1991); *ibid.*, **29**, 441 (1996).
5. M. Nishio and M. Hirota, *Tetrahedron*, **45**, 7201 (1989).
6. Y. Kodama, K. Nishihata, M. Nishio, and N. Nakagawa, *Tetrahedron Lett.*, 2105 (1977); M. Nishio, *Kagaku no Ryoiki*, **31**, 998 (1977); *ibid.*, **33**, 422 (1979).
7. Y. Kodama, K. Nishihata, S. Zushi, M. Nishio, J. Uzawa, K. Sakamoto, and H. Iwamura, *Bull. Chem. Soc. Jpn.*, **52**, 2661 (1979); Y. Kodama, S. Zushi, K. Nishihata, M. Nishio, and J. Uzawa, *J. Chem. Soc., Perkin 2*, 1306 (1980); J. Uzawa, S. Zushi, Y. Kodama, Y. Fukuda, K. Nishihata, K. Umemura, M. Nishio, and M. Hirota, *Bull. Chem. Soc. Jpn.*, **53**, 3623; S. Zushi, Y. Kodama, K. Nishihata, K. Umemura, M. Nishio, J. Uzawa, and M. Hirota, *ibid.*, **53**, 3631 (1980).
8. D. B. Amabilino, P. R. Ashton, J. F. Stoddart, S. Menzer, and D. J. Williams, *J. Chem. Soc., Chem. Commun.*, 2475 (1994).
9. P. R. Ashton, D. Philp, N. Spencer, J. F. Stoddart, and D. J. Williams, *J. Chem. Soc., Chem. Commun.*, 181 (1994).

10. D. Armspach, P. R. Ashton, R. Ballardini, V. Balzani, A. Godi, C. P. Moore, L. Prodi, N. Spencer, J. F. Stoddart, M. S. Tolley, T. J. Wear, and D. J. Williams, *Chem. Eur. J.*, **1**, 33 (1995).

11. M. Tamres, *J. Am. Chem. Soc.*, **74**, 3375 (1952).

12. C. M. Huggins and G. C. Pimentel, *J. Phys. Chem.*, **60**, 1615 (1956).

13. L. W. Reeves and W. G. Schneider, *Can. J. Chem.*, **35**, 251 (1957).

14. N. Nakagawa and S. Fujuwara, *Bull. Chem. Soc. Jpn.*, **33**, 1634 (1960).

15. R. E. Richards and J. F. Hatton, *Trans. Faraday Soc.*, **57**, 28 (1961).

16. J. W. Emsley, J. Feeney, and L. H. Sutcliffe, *High Resolution Nuclear Magnetic Resonance Spectroscopy*, Pergamon, Oxford, 1966, pp. 849–850.

17. T. P. Schaefer and W. G. Schneider, *J. Chem. Phys.*, **32**, 1218 (1960); A. D. Buckingham, T. P. Schaefer, and W. G. Schneider, *ibid.*, **32**, 1227 (1960).

18. N. Nakagawa, *Nippon Kagaku Zasshi*, **82**, 141 (1961); N. Nakagawa and S. Fujiwara, *Bull. Chem. Soc. Jpn.*, **34**, 143 (1961).

19. H. Shimizu, *Nippon Kagaku Zasshi*, **81**, 1025 (1960).

20. Z. Yoshida and E. Osawa, *Nippon Kagaku Zasshi*, **87**, 509 (1966); *Bull. Chem. Soc. Jpn.*, **38**, 140 (1965).

21. Y. Iitaka, Y. Kodama, K. Nishihata, and M. Nishio, *Chem. Commun.*, 384 (1974); Y. Kodama, K. Nishihata, M. Nishio, and Y. Iitaka, *J. Chem. Soc.*, *Perkin 2*, 1490 (1976).

22. M. Hirota, T. Sekiya, K. Abe, H. Tashiro, M. Karatsu, M. Nishio, and E. Osawa, *Tetrahedron*, **39**, 3091 (1983).

23. T. Aoyama, O. Matsuoka, and N. Nakagawa, *Chem. Phys. Lett.*, **67**, 508 (1979).

24. T. Takagi, A. Tanaka, S. Matsuo, H. Maezaki, M. Tani, H. Fujiwara, and Y. Sasaki, *J. Chem. Soc.*, *Perkin 2*, 1015 (1987).

25. S. Zushi, Y. Kodama, K. Nishihata, K. Umemura, M. Nishio, J. Uzawa, and M. Hirota, *Bull. Chem. Soc. Jpn.*, **54**, 2113 (1981); M. Nishio, *Kagaku no Ryoiki*, **37**, 243 (1983).

26. S. Araki, T. Seki, K. Sakakibara, M. Hirota, Y. Kodama, and M. Nishio, *Tetrahedron: Asym.*, **4**, 555 (1993).

27. H. Okawa, Y. Numata, A. Mio, and S. Kida, *Bull. Chem. Soc. Jpn.*, **53**, 2248 (1980); M. Nakamura, H. Okawa, and S. Kida, *Chem. Lett.*, 547 (1981); H. Okawa, K. Ueda, and S. Kida, *Inorg. Chem.*, **21**, 1594 (1982); H. Okawa and S. Kida, *Kagaku no Ryoiki*, **37**, 276 (1983); H. Okawa, *Coord. Chem. Rev.*, **92**, 1 (1988).

28. R. Ungaro, A. Pochini, G. D. Andreetti, and V. Sangermano, *J. Chem. Soc.*, *Perkin 2*, 1979 (1984); R. Ungaro, A. Pochini, G. D. Andreetti, and P. Domiano, *ibid.*, 197 (1985).

29. G. D. Andreetti, A. Pochini, and R. Ungaro, *J. Chem. Soc.*, *Chem. Commun.*, 1005 (1979).

30. The coordinates were kindly provided by Professor R. Ungaro.
31. G. D. Andreetti, O. Ori, F. Ugozzoli, A. Alfieri, A. Pochini, and R. Ungaro, *J. Incl. Phenom.*, **6**, 523 (1988).
32. M. Nishio, Y. Umezawa, M. Hirota, and Y. Takeuchi, *Tetrahedron*, **51**, 8665 (1995).

CHAPTER 2

EVIDENCE AND METHODS OF DETECTION

Evidence for a weak molecular interaction can be obtained from various types of measurements, classical as well as modern spectrometric means. Infrared (IR) spectra[1] reveal the specific involvement of hydrogen atoms as the perturbation on their vibrational modes in the molecular complex, and often give evidence about the putative XH/π bonding. In the early days, it was the most popular method among many spectroscopic means for detecting weak molecular forces such as hydrogen bonding. Nuclear magnetic resonance (NMR) spectroscopy[2] reveals the electronic environment of the hydrogen atom and provides the most useful means of detecting weak interactions. A high- or low-field shift in proton resonance demonstrates that the hydrogen is in a specific orientation relative to the π-electron system, thus verifying the XH to be complexed with the π-system. Optical rotatory dispersion[3] (ORD) or circular dichroism (CD) spectrometry, if appropriately used, may serve as a reliable tool for studying the nature of the interaction. Dipole moment[4] and molecular jet spectroscopy[5] show the presumed conformation of molecules and provide indirect evidence. Equilibrium constants in the complex formation also give evidence for the presence of weak molecular forces. Other techniques include classical methods such as thermochemical measurements.

The distance between atoms is an important criterion regarding the formation of a bond. A putative weak bond can thus be detected by its deviation from the normal van der Waals distance. The necessary information in searches of such anomalous atomic distances is obtained by X-ray or neutron diffraction studies. However, their application is limited to the interactions within crystals. Crystallographic data, when combined with the appropriate spectral and other data, may give the most reliable conclusion concerning the participation of weak intermolecular forces.

2.1. INFRARED SPECTROSCOPY

Intramolecular CH/π interactions in solutions could readily be detected by NMR spectroscopy. However, the CH/π-interacted conformers are short lived and we can observe only the average behavior of free and CH/π-interacted conformers by NMR.

In contrast, IR spectroscopy allows us to observe them individually. Generally speaking, a conformer has its own characteristic modes of vibration, which are different in some respect from any other conformer, and can be chosen for the quantitative determination of each conformer. Thus, every stable conformer (whose lifetime is longer than ca. 10^{-10} sec) shows its own absorption bands, distinct from those of other conformers. This implies that the free and the interacted species in the hydrogen bond and other weakly interacting systems can be observed distinctly by IR but not by NMR. Therefore, IR spectroscopy was employed as the method to study hydrogen bonding and other weak interactions.

Since CH/π interactions are similar in many respects to hydrogen bonding, discussions on their IR spectroscopic behaviors must start from a comparison with hydrogen bonding; IR spectra of hydrogen-bonded systems have been studied extensively. Assignment of vibrational modes and the effect of hydrogen bonding has been established unambiguously.[6] Let us consider a hydrogen-bonded system $R-X-H \cdots Y-R'$, where X and Y are electronegative atoms such as O, N, S or halogen. Hydrogen bonding perturbs several vibrational modes of these molecules. Several types of these vibrational modes are illustrated in Figure 2.1.

In weakly hydrogen-bonded systems, including CH/π-interacted ones, only the X–H stretching absorptions have been employed to detect its participation and to quantitatively determine the amount of the hydrogen-bonded species. The effect of hydrogen bonding on the other modes of vibration is often subtle and can be obscured by other

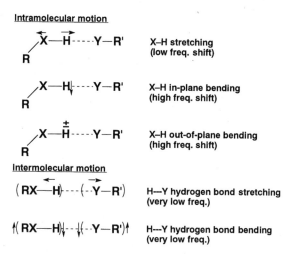

Figure 2.1. Vibrational modes of R–X–H $\cdots$ Y–R' hydrogen bond that may be used as the probe for hydrogen bonding.

effects. In the case of a CH/π interaction, it is, however, difficult to assign the C–H stretching absorption bands of the free and the π-bonded species. Absorptions in the C–H stretching region are complicated and many absorptions overlap;[7] this prevents the selection of any of the appropriate key bands. To overcome this difficulty, ^{2}H-isotopomers are conveniently used because (1) C–D stretching absorption moves toward the lower frequency from the C–H stretch region owing to the mass effect, which allows us to differentiate the key bands from the other C–H absorptions and (2) no interfering absorption bands appear in the 2400–2000 cm^{-1} region where the C–D absorption appears.

2.1.1. Frequency Shifts and Intensity Enhancements of C–H Stretching Bands

In a hydrogen-bonding system XH/Y (where X–H is a hydrogen donor and Y is a hydrogen acceptor), the X–H stretching absorption band of the hydrogen-bonded species shifts toward the lower frequency in comparison with the free species, and at the same time broadens out. The trend becomes more dominant as the hydrogen bond becomes stronger. The frequency shift can be observed regardless of whether the hydrogen bonding interaction occurs intra- or

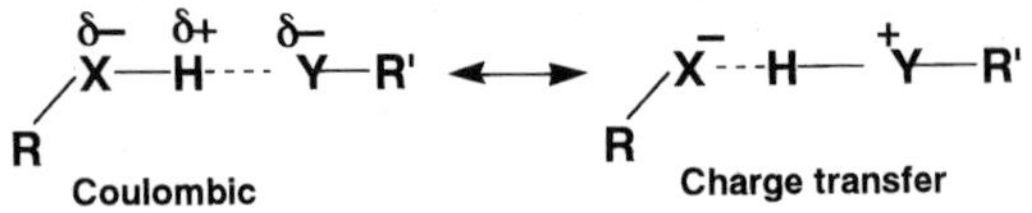

Figure 2.2. Low-frequency shift of X–H stretching band due to the charge transfer contribution to hydrogen bond.

intermolecularly. The shift is due to the weakening of the X–H bond caused by hydrogen bond formation, which can be interpreted as the charge transfer (see Fig. 2.2, the canonical structure **1**).

In sharp contrast, intramolecular hydrogen bonding involving C–H bond CH/X (X = O, N, S, etc.) accompanies a high-frequency shift of the C–H absorption band.[8] The low-frequency shift is rather small even in the intermolecular CH/X interaction.[9] In a CH/π interaction, the C–H stretching band behaves quite similarly to the CH/X hydrogen bonding—a small low-frequency shift in the intermolecular case and a large higher-frequency shift in the intramolecular case.

An example of intermolecular CD/π interaction (CDCl$_3$-arene) is given in Table 2.1.[10] Similar Δv-values for the OH/π interaction are 6–7 times as large as the CH/π cases: 38 cm^{-1} for CH$_3$OH–C$_6$H$_6$ and 75 cm^{-1} for CH$_3$OH–C$_6$Me$_6$.

The terminal CH groups of acetylenes were also shown to behave as proton donors by measuring their infrared C–H stretching bands. The dimer formation of RC≡CH[11] and the association of benzoylacetylenes[12] with aromatic solvents were attributed to the formation of

TABLE 2.1. Frequency Shifts (cm^{-1}) in the C–D Absorption Band of CDCl$_3$ in Neat Aromatic Compounds

Solvent	v_{CD} (free)	$\Delta v\ (= v_{free} - v_{interacted})$
Gas phase		—
CCl$_4$	2252	0
C$_6$H$_6$	2247.5	4.5
m-C$_6$H$_4$Me$_2$	2247	5
C$_6$Me$_6$	2244	8
C$_6$H$_5$Cl	2249	3
C$_6$H$_5$CH$_3$	2247	5
1,3,5-C$_6$H$_3$Me$_3$	2246.5	5.5

intermolecular CH/π interaction on the basis of the frequency shifts of the acetylenic CH bands. Huggins and Pimentel[13] reported that the intensity of the C–D stretching band of chloroform-d in benzene and mesitylene showed significant enhancement, which indicates the formation of a CD/π complex.

Intramolecular interaction of the CH group to a π-group usually causes a high-frequency shift of v_{C-H} band. Similar high-frequency shifts are often observed when the C–H bond is compressed by steric congestion.[14] Typical examples of steric compression effect are given in Figure 2.3a.

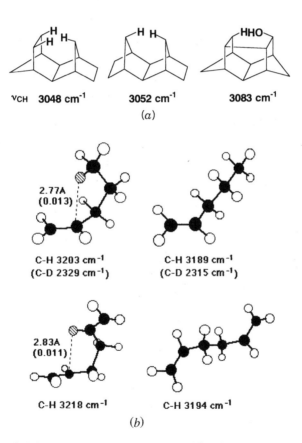

Figure 2.3. High frequency shift of C–H stretching bands. (a) Unusually high frequency CH absorptions due to steric compression of C–H bonds. (b) Ab initio calculations on C–H/π proximate conformers. (Non-bonded H ··· C overlap populations are given in parentheses.)

TABLE 2.2. C–D Stretching Frequencies of Ketones R–CO–CDMe$_2$

R	ν_{CD} (cm^{-1})	R	ν_{CD} (cm^{-1})
Ph	2191, 2156	CMe$_3$	2195, 2182
PhCHMe	2177, 2136	CHMe$_2$	2182, 2154
PhCH$_2$	2153, 2134	CH$_2$Me	2168, 2130
PhCH$_2$CH$_2$	2157, 2130	CH$_3$	2167, 2130

The C–D stretching frequencies of several 2-propyl-2-d ketones are given in Table 2.2. As is easily understood from the table, the C–D frequencies of alkyl ketones (right column) tend to increase as the steric congestion in the molecules increases. Similar high-frequency shifts are observed with aryl- or aralkyl-substituted derivatives (R = Ph or aralkyl) in the left column. The band at the higher frequency (2177 cm^{-1} for R = CHMe) is assigned to the CH/π-interacted conformer on the basis of the electronic effect by the p-substitued groups (Section 2.1.2). Thus, the high-frequency shift in the intramolecularly CH/π interacted molecules can be ascribed to the steric compression of a CH group in the interacted conformers. However, the fact seemed contradictory to the hydrogen-bond-like delocalization effect and needed to be pursued further.

Our recent MP2/6-31 G* calculations gave a support on this assignment. The vibrational frequency of the CH bond opposing to the π-orbital were calculated with the CH/π proximate conformers of 1-pentene and 1,5-hexadiene (Fig. 2.3b). The stretching frequencies of C–H groups proximate to π-systems were calculated to be higher than those of the stretched conformer. Simultaneously the non-bonded bond populations involving these CH become considerably positive. This fact can be crucial evidence for the above assignment of C–H stretching bands.

2.1.2. Determination of the Thermodynamic Quantities of CH/π Interaction by IR Spectroscopy

Infrared Spectroscopy provides a very straightforward method of determining the equilibrium constant from the ratio of the amounts of the free and the interacted species by Eq. 2.1.

$$K_{int} = [\text{interacted}]/[\text{free}] = a(A_i/A_f) \qquad (2.1)$$

In this equation, A_i and A_f are the integrated intensities of the key bands for interacted and free species, respectively, and a is a constant representing the ratio of the intrinsic molar intensities of the two bands.

An example in which deuterium-labeled derivatives was applied is given below.[15] The C–D stretching absorptions of the free and the CD/π interacted rotamers of a series of substituted 1-phenylethyl isopropyl-1-d ketones **2** were measured, from which the relative formation constants ($K_{rel} = [\mathbf{2}_{int}]/[\mathbf{2}_{free}]$) were determined as given in Table 2.3. The temperature dependence of the intensities allows us to determine the formation constant as a function of temperature, from which the enthalpies and the entropies of the complex formation can be obtained by the van't Hoff equation (Eq. 2.3) based on Gibbs–Helmholtz equation (Eq. 2.2).

$$(\partial \ln K_p / \partial T)_p = \Delta H / R T^2 \tag{2.2}$$

$$\ln K = -(\Delta H / R) T^{-1} + (\Delta S / R) \tag{2.3}$$

$\mathbf{2}_{int}$ $\mathbf{2}_{free}$

The slope of the K_{rel} versus σ plot (Fig. 2.4) is negative and the formation constant (K_{rel}) tends to increase as the substituent X becomes more electron-donating. ^{2}H NMR chemical shifts of these ketones move toward the higher fields as the substituent becomes more electron-donating. The high-field shift can be attributed to the increase in the population of the CH/π-interacted conformation in which the CH group above the aromatic nucleus receives a diamagnetic anisotropy effect by the ring current.

Another example is benzyl t-butyl ketone. The carbonyl absorption band of this ketone is asymmetric, having a maximum at $1716.5 \, \text{cm}^{-1}$ and a shoulder at $1704.5 \, \text{cm}^{-1}$ (Fig. 2.5). The absorption band can be separated into two component peaks. The stronger band at the higher frequency ($1716.5 \, \text{cm}^{-1}$) is assigned to the Ph/Bu^t anti conformer on the basis of (1) the solvent effect on the relative intensities showing that it is the more polar conformer (calculated dipole moments for anti and

TABLE 2.3. Relative Formation Constants and CD/π Interaction Enthalpies from IR Data (in CCl₄) and ²H-NMR Data (in CCl₄) of p-XC₆H₄CH(CH₃)COCD(CH₃)₂

X	Infrared v_{CD}			^{2}H-NMR $(\delta/\text{ppm})^a$
	v_{int}/cm^{-1}	v_{free}/cm^{-1}	$\varepsilon_{int}/\varepsilon_{free}$ (K_{rel})	
NO$_2$	2174	2135	1.10	− 5.30
Br	2174	2135	1.29	− 5.36
Cl	2175	2136	1.41	− 5.36
H	2177	2136	1.59	− 5.37
C$_2$H$_5$	2176	2136	1.54	− 5.37
CH$_3$	2176	2136	1.59	− 5.38
NH$_2$	2175	2135	1.64	—

^{a} ^{2}H chemical shifts given by parts per million downfield from external CDCl$_3$ standard.

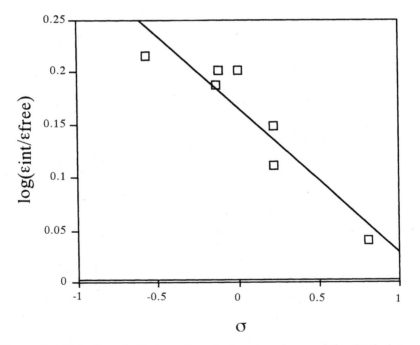

Figure 2.4. Substituent effect on the relative abundance of the CH/π interacted species in p-XC₆H₄CH(CH₃)COCD(CH₃)₂.

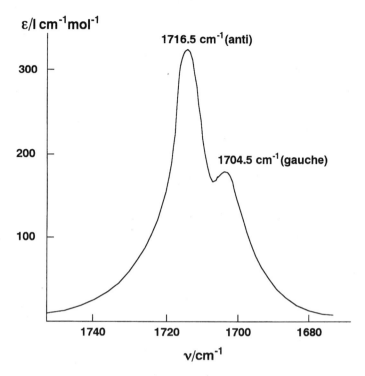

Figure 2.5. Carbonyl stretching absorption bands of anti (free) and gauche (CH/π interacted) conformers of benzyl t-butyl ketone.

gauche are 2.81 and 1.35 D, respectively), and (2) the general trend seen in the high-frequency shifts of carbonyl groups proximate to aromatic π-electrons due to electrostatic repulsion. There are two more pieces of evidence: variable temperature measurements of the intensities of these two bands gave a good linear $\log K$ versus T^{-1} relation ($r = 0.99$), from which $\Delta H = 0.25$ kcal mol^{-1} (in CCl$_4$ solution)[16] was obtained by using Eq. 2.2, which agrees well with the calculated steric energy difference of MM2 of 0.18 kcal mol^{-1}.

2.2. NUCLEAR MAGNETIC RESONANCE SPECTROSCOPY

The ordinary hydrogen bond is known to shift the proton resonance to the lower magnetic field; the low-field shift has been rationalized as the effects of the electric field and the magnetic anisotropy caused by the lone-pair electrons on the hydrogen-accepting atom. The proton magnetic resonance signal of $CHCl_3$, however, is known to shift to the higher field in aromatic solvents. The phenomenon has been ascribed to the diamagnetic anisotropy effects of the π-electrons in the aromatic ring. This serves to indicate the presence of a specific type of interaction between a CH and a π-system.

2.2.1. Rotamer Populations of Triptycene Derivatives—An Example of Long-Lived Rotamers

Ōki et al.[17] studied the effect of substituents on rotamer populations of a series of 9-benzyltriptycene **3**. Rotational barriers separating the conformers **3a** and **3b** (Fig. 2.6) are high enough to give individual NMR signals for synclinal (sc) and antiperiplanar (ap) conformers and the equilibrium constant was calculated as the ratio of their intensities.

Table 2.4 summarizes the results. Introduction of a group (X) into the benzyl group affected the sc versus ap ratio; the highest value was recorded for X = NMe_2, while the lowest was obtained for X = NO_2. The opposite was seen on varying the substituent Y, which is located para to the methyl group of the benzeno ring; the increase in the electron-withdrawal properties of Y increased the sc/ap ratio. The trend can be rationalized only if we assume that the CH/π interaction is playing a role.

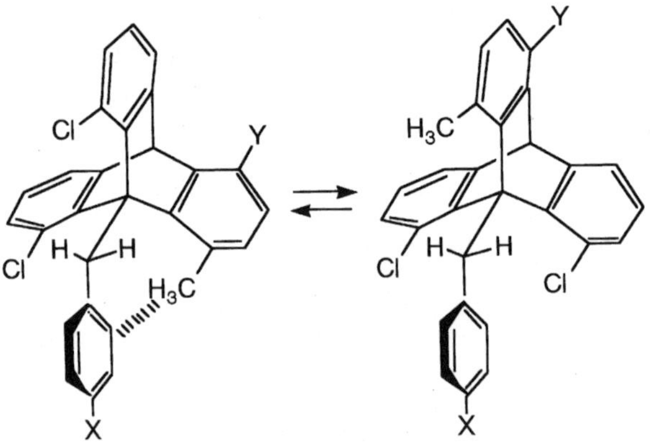

Figure 2.6. Rotamers of 9-benzyltriptycene.

TABLE 2.4. **Equilibrium Constants of the Rotational Isomers (sc/ap) of the Triptycene Derivatives 3 at 54°C in CDCl$_3$**

X/Y	CH$_3$	H	COOCH$_3$	CN
N(CH$_3$)$_2$		2.87		
H	2.30 (1.08)	2.22 (1.01)	4.09 (0.98)^b	3.42 (1.05)
NO$_2$		1.52		

^{a} In parentheses are differences in enthalpy (kcal mol^{-1}) between the rotamers estimated by the MM2 method.
^{b} For X = COCH$_3$.

2.2.2. Chemical Shift Anomalies or ASIS in Proton NMR

Reeves and Schneider[18] examined the association of chloroform with aromatic and olefinic solvents by ^{1}H NMR spectroscopy. The ^{1}H signal of chloroform shifted upfield significantly in benzene, toluene, or mesitylene. This implies the presence of a specific interaction between chloroform and the solvents. This type of shift of δ_{CH} in aromatic solvents has been extensively studied and is known as the aromatic solvent-induced shift (ASIS),[19] as shown in the Bovey model (Fig. 2.7).

This leads to the conclusion that the hydrogen atom in chloroform should be located, on average, close to and just above the plane of the aromatic ring **4**.

4

An appreciable upfield shift of the ethinyl proton of phenylacetylene occurs on dilution with benzene, and was attributed to the complex formation of the solute with the solvent benzene.[20] The effect of benzene on proton resonances of a variety of compounds bearing a methyl group was further pursued.[21] The difference ($\Delta\delta = \delta_b - \delta_c$) in the ^{1}H chemical shift of the methyl group in benzene (δ_b) in reference to the shifts in carbon tetrachloride (δ_c) was taken as a measure of complex formation. The $\Delta\delta$ (Table 2.5) was related to the CH acidity of the methyl group.

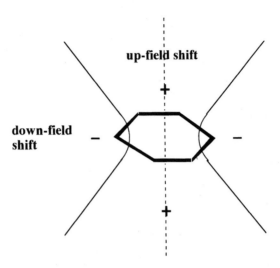

Figure 2.7. The Bovey model for the magnetic anisotropy of benzene.

TABLE 2.5. Methyl Proton Resonances in Carbon Tetrachloride (δ_c) and Benzene (δ_b)

	δ_c^a	δ_b^a	$\Delta\delta$ ($\delta_c - \delta_b$)
Acetone	0.68	0.24	0.43
Acetophenone	1.12	0.72	0.40
Acetic anhydride	0.74	0.25	0.49
Methyl iodide	0.72	0.09	0.63
Acetonitrile	0.53	− 0.43	0.97
Nitromethane	2.88	1.75	1.13

[a] Parts per million downfield from internal cyclohexane.

A good linear relation was obtained between $\Delta\delta$ and the Hammett' σ for p- and m-substituted toluenes (X-$C_6H_4CH_3$, $X = NO_2$, CN, Cl, H, Me, OH, NH_2, Fig. 2.8).

The ASIS has been treated quantitatively in order to obtain information concerning thermodynamic quantities of CH/π complexes. Even though we cannot observe the NMR signals of the free and the interacted species separately, the equilibrium (formation) constants can easily be determined by the measurement of the NMR chemical shifts when the exact chemical shift values of the free and the

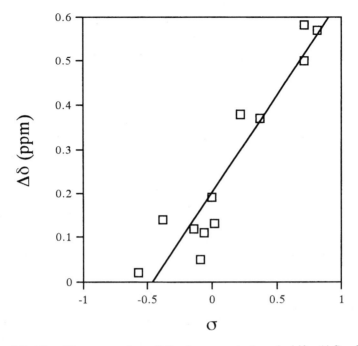

Figure 2.8. The Hammett plot of the benzene-induced shifts ($\Delta\delta$) of CH_3 signals for p- and m-substituted toluenes, X-$C_6H_4CH_3$.

interacted (complexed) species (δ_0 and δ_c, respectively) are known. The chemical shift δ of the equilibrium mixture is expressed as the weighted mean of those of the free and the CH/π interacted species:

$$\delta = \delta_0(1 - p) + \delta_c p \qquad (2.4)$$

where p is the mole fraction of the interacted species. This method has been applied to determine the K_{int} for intermolecular CH/π systems.

The method for the determination of the formation constant by NMR spectroscopy is well established and has long been known.[22] If we assume that only the 1:1 complex (**C**) is formed by the CH/π interaction, the formation constant K_{int} can be evaluated by a combination of Eqs. 2.4 and 2.5 from the high-field shift of δ_D.

$$K_{\text{int}} = pc_D/[c_D(1 - p)(c_A - pc_D)] \qquad (2.5)$$

where c_D and c_A are the initial stoichiometric concentrations of the CH donor (**D**) and the π-base (acceptor, **A**), respectively. The mole fraction p can be calculated by Eq. 2.4 from the observed chemical shift δ of the CH donor in the CH/π interacted system, if we could estimate the intrinsic chemical shift of the CH/π interacted complex (δ_c). The δ_c values were evaluated experimentally by an extrapolation method. The formation constants at various temperatures gave the enthalpies and entropies of formation by use of the van't Hoff plot, from which ΔH and ΔS can be obtained by Eq. 2.3.[23]

2.2.3. Lanthanide-Induced NMR Chemical Shifts

Lanthanide-induced shift (LIS) has been applied to determine the conformations of organic molecules bearing a functional group capable of forming complexes with the lanthanide shift reagent (LSR).[24] The method is based on the McConnell–Robertson relationship (Eq. 2.6),[25] which describes the dependency of LIS on the distance (r) and the direction (κ) from the lanthanide ion (Ln^{3+}):

$$(\text{LIS})_i = K(3\cos^2\kappa_i - 1)/r_i^3 \qquad (2.6)$$

where r_i is the length of the vector joining the paramagnetic center (lanthanide ion, Ln^{3+}) and the ith nucleus (N_i), and κ_i is the angle between this vector and the principal magnetic axis of Ln^{3+} (Fig. 2.9).[26]

Studies on the conformations of a series of compounds with the general structure PhCH(Me)–X–R and PhCH$_2$–X–R[27] have suggested that the conformations bearing a synclinal R/Ph relationship predominate in the rotameric equilibria in this series of compounds. Thus, Kodama et al. wrote a program[28] to calculate LIS from the NMR signals of flexible molecules such as $C_6H_5CH(Me)$–X–R,

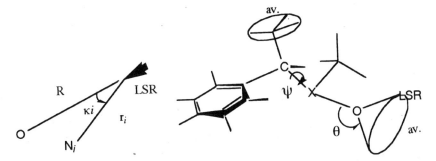

Figure 2.9. Procedure for the simulation of LIS.

according to Eq. 2.6. The approximations are (1) the contact shift was neglected and (2) the contribution from the nonaxial term of the dipolar field was neglected. Conformation of the molecule was described by a single set of coordinates and the values for the nuclei of α-methyl, alkyl, and phenyl groups were computed with various geometries and then averaged. As to the geometry of the LSR-substrate complex, the best-fit X–O–Ln angle (θ, 120–140°) and O–Ln distance (R, 3.2–3.4 Å) were searched by a trial and error method (Fig. 2.9).

Structural parameters and LIS data were then input for the calculation, and the conformations of the molecules were varied, by rotating stepwise the C(phenyl)–C–X–O torsional angle (ψ), to examine the agreement of the computed values with the experimental ones of LIS. The Hamilton agreement factor (AF, Eq. 2.7)[29] was used to assess the agreement between the calculated and the observed LIS data.

$$AF = [\Sigma(LIS^{obsd} - LIS^{calcd})^2 / \Sigma(LIS^{obsd})^2]^{1/2} \qquad (2.7)$$

Figure 2.10 gives AF profiles plotted against the O/Ph torsional angle (ψ) for $C_6H_5CH(Me)$–SO–R **5** ($R = Bu^t$). A distinct AF minimum has been observed at ψ around 190° and 310°, respectively, for *threo*- and *erythro*-sulfoxides.[30] The LIS results (Fig. 2.11) agree with those obtained by X-ray crystallography (Section 2.4.1). It therefore follows that the torsional angles as obtained above reflect the geometry of the molecules in solution, namely, the *t*-butyl group prefers to lie close to the phenyl group, in contrast to the generally held idea that the bulky group (*t*-Bu and Ph) should be separated as far as possible in the stable conformations.

Similar AF/ψ profiles were obtained for lower alkyl homologs **5** ($R = Me$, Et, Pr^i).[31] Further, the generality of the phenomenon

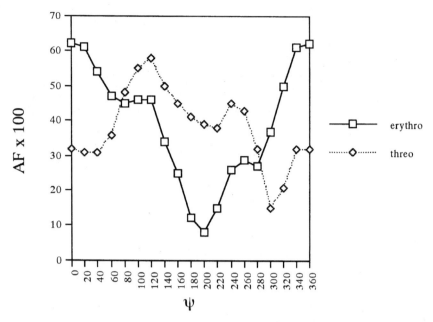

Figure 2.10. AF profiles plotted against the O/Ph torsional angle (ψ) for *(RR/SS) threo* and *(RS/SR) erythro*-1-phenylethyl *t*-butyl sulfoxides.

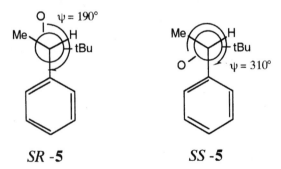

Figure 2.11. Solution conformations of a diastereomeric pair of sulfoxides **5** (R = Bu^t).

was explored for molecules with related structures.[32] Figure 2.12 illustrates similar treatments of *threo*-C$_6$H$_5$CH(Me)CHOH-R **6** (Fig. 2.13). Rotamer **A** (sc-R/Ph, ap-R/Me) is the most preferred throughout the series.[33]

The LIS results above for **5** and **6** are in accord with other methods: X-ray crystallography, ^{1}H and ^{13}C chemical shifts, NMR

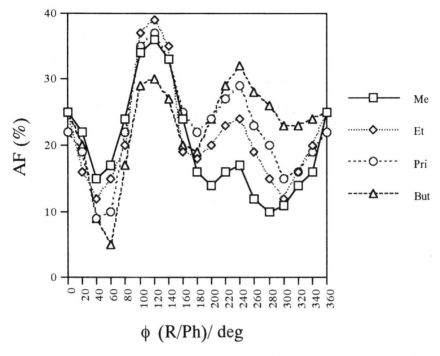

Figure 2.12. AF profiles plotted against the R/Ph torsional angle (ϕ) for 1-phenylethyl alkyl carbinols **6** with the *threo* configuration.

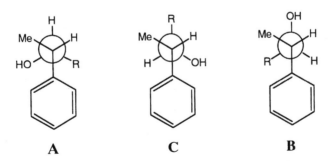

Figure 2.13. *Threo*-C_6H_5CH(Me) CHOH-R **6**.

spin-coupling data, circular dichroism data, and dipole moment measurements. The experimental results are internally consistent and point to the conclusion that the alkyl/phenyl sc conformers are generally preferred. A similar conclusion has been obtained for compounds bearing a π-group other than C_6H_5, such as $CH{=}CH_2$ or $CMe{=}CH_2$.[34]

2.2.4. Nuclear Overhauser Effect

The nuclear Overhauser effect (NOE)[35] is a convenient method to find out the nuclei located proximate to each other. The magnitude of NOE is dependent on the distance between the two nuclei, but not on the bond connectivity. The NOE enhancement is closely related to the dipole–dipole relaxation interaction, which tends to decrease as a function of r^{-6} (where r is the internuclear distance). The magnitude of enhancement decreases quickly as the internuclear distance increases (i.e., a very short-range effect). When the nucleus S was saturated by irradiation and the nucleus I was observed, the enhancement $f_I(S)$ can be expressed by Eq. 2.8.

$$f_I(S) = K(\gamma_S/\gamma_I)r^{-6} \qquad (2.8)$$

where γ_S and γ_I are the gyromagnetic ratios of the nuclei S and I, respectively. Thus, NOE is a powerful method to detect the "through-space" proximity of two atoms which may be separated by several bonds. In the CH/π-interacted conformation, the CH proton is located close enough to one or several aromatic C or H atoms to show a considerable NOE enhancement.

Many CH/π interacted compounds showed NOE enhancements of the aromatic H signal(s) when the donor H atom was irradiated. It is difficult to estimate the absolute value for the population of the CH/π interacted conformer by an NOE experiment. However, comparison of the magnitudes of NOE through a series of similar compounds allows us to estimate trends.[36] Intramolecular CH/π interaction depends critically on the conformation and can persist only in CH/π proximate conformers. When the CH/π proximate conformer is realized, a significant part may be involved in the CH/π interaction if it leads to some stabilization of the conformer. The NOE enhancements of aromatic protons induced by irradiation of the donor CH proton are given in Table 2.6 for several series of compounds capable of forming intramolecular CH/π interaction.

As is easily seen from Table 2.6, the NOE enhancement $[f_{H_{ar}}(H_{donor})]$ becomes stronger as the substituent becomes more electron-donating.

TABLE 2.6. The NOE Enhancements [$f_{H_{ar}}(H_{CH})$/%] of Aromatic Protons[37] Induced by the Irradiation of CH Donor Proton (italicized)

Structures **7**, **8**, **9**

7			8			9		
X			X			X		
CH$_3$O	H	Br	CH$_3$O	H	NO$_2$	CH$_3$O	H	NO$_2$
7.4	4.6	4.4	3.2	2.9	1.5	0.5	0.7	0.4

Structures **10**, **11**

10					11		
X					X		
CH$_3$O	CH$_3$	H	Cl	NO$_2$	CH$_3$	H	Cl
4.2	3.4	3.0	2.8	2.2	4.2	3.8	3.2

The trend is most clearly revealed in a series of benzyl formates **10**, of which $f_{H_{ar}}(H_{CHO})$ versus σ plot are shown in Figure 2.14. Their NOE enhancements [$f_{H_{ar}}(H_{CHO})$] are dependent on the electronic properties of p-substituents in contrast to those of benzyl protons [$f_{H\alpha}(H_{CHO})$], which are almost conformation-(hence, substituent)-independent throughtout the series [$f_{H\alpha}(H_{CHO}) = 1.8\%$ for X = CH$_3$O and 1.9% for X = NO$_2$]. The $f_{HAr}(H_{CHO})/f_{H\alpha}(H_{CHO})$ ratio increases as the substituent becomes more electron-donating, suggesting that the CH/π contiguous conformer **10a** (Fig. 2.15) should become more favorable in this order. A similar but slightly more significant trend of the substituent effect was observed with substituted 1-phenylethyl formates **12**.

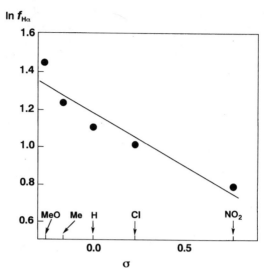

Figure 2.14. Logarithm of NOE versus σ plot for *p*-substituted benzyl formates **10**.

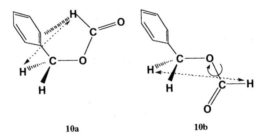

Figure 2.15. The CH/π contiguous **10a** and stretched conformers **10b** of benzyl formate. Average H_α/H_{formyl} distances are nearly the same between **10a** and **10b**.

If we compare the NOE enhancements $[f_{H\alpha}(H_{CHO})]$ of the series of isopropyl compounds $[C_6H_5CH_2YCH(CH_3)_2]$ capable of forming five-membered intramolecular CH/π interactions (Table 2.7),[38] the CH/π interacted conformer is suggested to increase in the order **13** (Y = CO) < **14** (Y = NCH₃) < **7** (Y = O). This sequence agrees with the increasing order of electronegativity of the connecting group Y, suggesting that more positively charged and, hence, acidic hydrogen atom forms stronger CH/π interactions with a π-base.

The formyl CH groups of the aldehyde **16** and the formate ester **10, 11** are assumed to be more acidic than the isopropyl CH of the ketone; this is reflected in their larger NOEs. Intramolecular CH/π

TABLE 2.7. NOE Enchancement $[f_{H_{arom}}(CH)]$ by Five-Membered CH/π Interaction^a

Compound	$f_{H_{arom}}(H_{donor})/\%$	
	R = H	R = CH$_3$
C$_6$H$_5$CH**R**C(=O)CH(CH$_3$)$_2$ **13**	1.4	
C$_6$H$_5$CH**R**N(CH$_3$)CH(CH$_3$)$_2$ **14**	3.7	4.8
C$_6$H$_5$CH**R**OCH(CH$_3$)$_2$ **7, 12**	4.6	5.5
C$_6$H$_5$C(=O)N(CH$_3$)CH(CH$_3$)$_2$ **15**	4.8	
C$_6$H$_5$CH**R**CH$_2$CH=O **16**	1.7	
C$_6$H$_5$CH**R**OCH=O **10, 11**	3.0	3.8

^{a} Donor H atoms are shown by the bold letters **H**.

interaction of these series of compounds are considerably enhanced by introducing an alkyl (usually methyl) substituent on the benzyl carbon atom. Thus, the 1-phenylethyl series (R = CH$_3$) is always more favorable than the benzyl series (R = H) in forming the CH/π interaction.

2.3. CIRCULAR DICHROISM

Molecular orbital calculations have been applied to the interpretation of the effect of CH/π interactions on the chiroptical properties. The charge-transfer interaction between the CH group and π-system should induce asymmetric deformation of the π-electron cloud when the CH group approaches from a direction other than that of the plane of symmetry. In this way, CH/π interaction can induce an additional optical rotatory power of the (π, π^*) transition of the π-system (ethylenic, dienic, and aromatic chromophores). This hypothesis has been confirmed by the calculations of the rotational strengths in some optically active CH/π interacted systems.

Rotational strengths R can be calculated by using the Rosenfeld equation (Eq. 2.9):[39]

$$R = \text{Im}\{\langle \Psi_0|\mu|\Psi_e\rangle\langle \Psi_e|m|\Psi_0\rangle\} \qquad (2.9)$$

Here, Im means imaginary part. Ψ_0 and Ψ_e are the wave functions of the ground and the excited states related to the transition in question, and μ and m are the electric and magnetic dipole moment operators, respectively. Equation 2.9 was developed into atomic orbital terms suitable for the application to semiempirical MOs, giving finally Eq. 2.10:

$$R = [-7.313 \times 10^3/(E_j - E_i)]\langle\phi_i|\nabla|\phi_j\rangle\langle\phi_j|r \times \nabla|\phi_i\rangle \quad (2.10)$$

where

$$\langle\phi_i|\nabla|\phi_j\rangle = \sum_r \sum_s c_{ir}c_{js}\langle\chi_i|\nabla|\chi_j\rangle$$

$$\langle\phi_j|r \times \nabla|\phi_i\rangle = \sum_r \sum_s c_{ir}c_{js}\langle\chi_j|r \times \nabla|\chi_i\rangle$$

2.3.1. CD Spectra of Olefins

Exomethylene steroids[40] show significant enhancement in the Cotton effect amplitude of a π–π^* transition when an axial methyl group is present allylic to the double bond. In such an arrangement, a CH/π interaction can occur. For instance, 4-methylene-5α-androstane **17** and 6-methylene-5α-androstane **18** have an axial methyl group capable of intramolecular CH/π interaction with exocyclic ethylenic π-bonds.

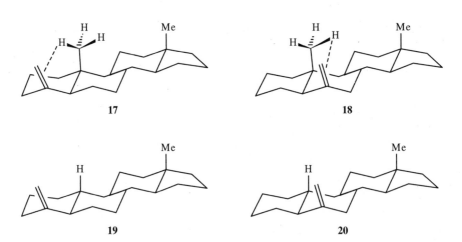

The rotational strengths of these compounds have in fact been found to be much greater than the molecules lacking the 19-axial methyl group in accordance with our expectation.[41] Let us compare 4-methylene-5α-androstane **17** with 4-methylene-5α-estrane **19**. Their geometries are expected to be similar, but the 19-methyl group in the

TABLE 2.8. Rotational Strengths (R) and CD Data ($\Delta\varepsilon$)[42] of Some Exo-Methylene Steroids

Compound	ΔE/eV	λ_{obs}/nm	$R/10^{40}$ esu	ΔR_{int}	$\Delta\varepsilon_{obs}$	$\Delta\Delta\varepsilon_{int}$
17	6.01	204.4	− 40.71	− 4.10	− 10.5	− 6.4
18	6.08	203.8	− 36.61		− 4.1	
19	6.05	205.0	+ 0.93	+ 12.10	+ 4.2	+ 4.5
20	6.07	204.1	− 11.17		− 0.3	

former may enter into through-space interaction with the 4-methylene group. The effect of CH/π interaction on the optical rotatory power should operate in a through-space manner, and can be measured as the difference between the rotational strengths of the corresponding andros-tanes (bearing 19-CH$_3$) and estranes (void of 19-CH$_3$). The calculated rotational strengths and the observed CD data[42] are given in Table 2.8.

The contribution by CH/π interaction (ΔP_{int}) can be assumed to be evaluated as the difference of R between the interacted and non-interacted systems; thus, by substracting the R of methylene–estrane (**18** or **20**) from that of corresponding methylene–androstane (**17** or **19**). The calculated rotational strengths R, as well as ΔR_{int}, agree qualitatively with the observed CD data ($\Delta\varepsilon_{obs}$ and $\Delta\Delta\varepsilon_{int}$), and the changes in $\Delta\varepsilon$ by the introduction of a 19-methyl group could be rationalized by taking into account the perturbation on the (π, π^*) transitions via a through-space CH/π interaction. Simple model calculations on the methane (as the CH-donor)/ethylene (as the π-base) supramolecular system have been carried out. A nonzero induced rotational strength was obtained when the CH bond was located off the symmetrical plane of the ethylene molecule. As both components are achiral, the calculated rotational strength of the π, π^* transition should be originated from the chiral distortion of the π-electron distribution by the interaction with a CH bond of the methane molecule.

2.4. X-RAY CRYSTALLOGRAPHY

Distance between atoms is a good criterion regarding the formation of a weak bond. The CH/π interaction, if present, must therefore be reflected in the crystallographic data, as deviations from the normal van der Waals distance.

2.4.1. Crystal Structures of a Diastereomeric Pair of Sulfoxides

Iitaka et al. determined the crystal structures of a diastereomeric pair of 1-(*p*-bromophenyl)ethyl *t*-butyl sulfoxide **5** (R = Bu^t, X = Br).[43] In both compounds, one of the methyls in the *t*-butyl group has been found to lie close and above the benzene ring (Fig. 2.16 and Table 2.9).

The distances between one of the methyl carbons (C_{Me^2}) in the *t*-butyl group and the aromatic *ipso* carbon (C^1) are very short; 3.24 and 3.32 Å, respectively, for the (*SR*)/(*RS*) and (*SS*)/(*RR*) diastereomer. These values are much shorter than the value expected from the van der Waals contact (ca. 3.7 Å: 2.0 Å for CH_3 and 1.7 Å for C_{sp^2}).

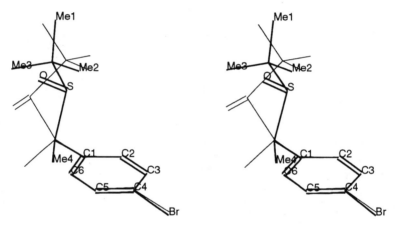

Figure 2.16. Stereo view of 1-(*p*-bromophenyl)ethyl *t*-butyl sulfoxides **5**. Thick lines: (*SR*) isomer; thin lines: (*RR*) isomer.

TABLE 2.9. C_{sp^3}/C_{sp^2} **Interatomic Distances of a Diastereomeric Pair of 1-(*p*-Bromophenyl)-ethyl *t*-Butyl Sulfoxides**

	(*SR*)/(*RS*) Å	(*SS*)/(*RR*) Å
C^1-C(Me^2)	3.24	3.32
C^2-C(Me^2)	3.68	3.61
C^6-C(Me^2)	3.61	3.74

2.4.2. Crystal Structures of Several Organic Compounds

Short H/C interatomic contacts are often found in a variety of molecules including natural organic compounds. For instance, the crystal structure of levopimaric acid[44] contains a H/C_{sp^2} distance much shorter ($C^{17}H/C^8$ 2.52 Å) than the sum of the van der Waals distances of the relevant atoms. A number of short interatomic contacts have also been noted[45] between a CH and a sp^2 carbon in crystallographic data of triterpenes such as lumisterol (2.74 Å: $C^{18}H/C^8$),[46] pyrocalciferol (2.75 Å: $C^{18}H/C^8$),[47] and isopyrocalciferol (2.57Å for $C^{18}H/C^8$).[48] The distance calculated by assuming van der Waals distances is between 2.9 and 3.1 Å (1.2.–1.4 Å for H and 1.7 Å for C_{sp^2}) for H/C_{sp^2}.

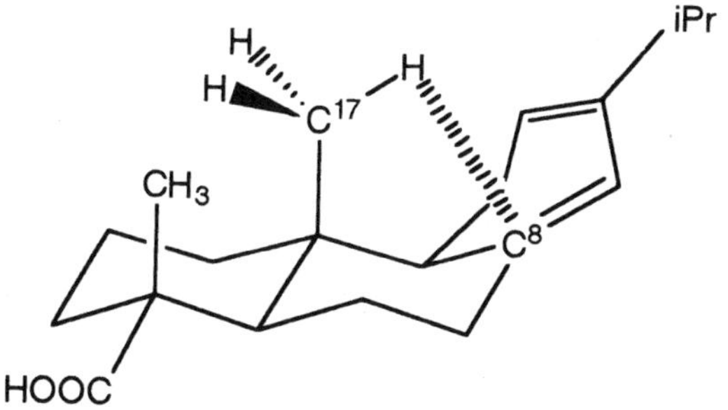

levopimaric acid

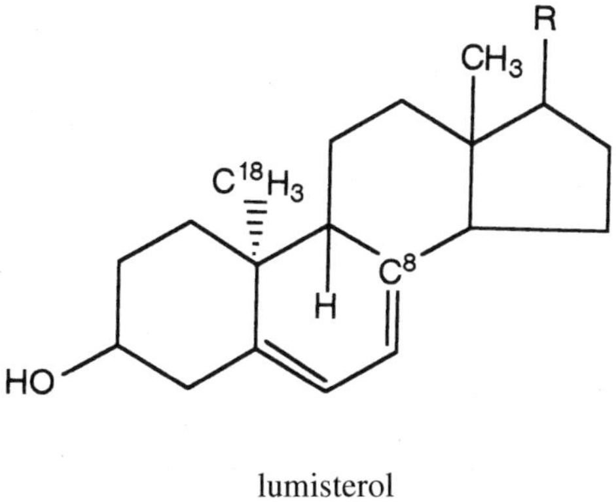

lumisterol

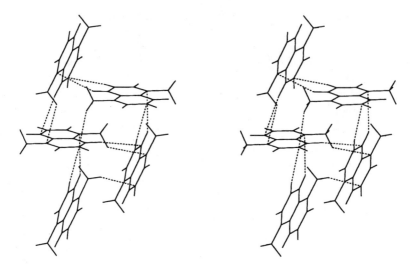

Figure 2.17. 1,5-dimethylnaphthalene (Cambridge Structural Database[49] refcode DIMNAP, stereo view). Dotted lines indicate CH/C_{sp^2} contacts.

Figure 2.17 shows CH/C_{sp^2} contacts ($< 2.9\,\text{Å}$) found in the crystal structure of 1,5-dimethylnaphthalene determined by neutron diffraction.[49] A molecule is shown to be surrounded by many other 1,5-dimethylnaphthalene molecules with CH/π interactions.

2.5. MISCELLANEOUS METHODS

Conformational analysis[50] of simple organic compounds may provide information, though indirect, for the occurrence of attractive interactions between interacting groups. Such information has been obtained by spectral means such as ORD/CD and molecular jet spectrometry, or dipole moment measurements. More information can be obtained from the substituent effect on a variety of experimental systems, such as stereoselectivity in coordination chemistry and substrate specificity of inclusion complexes.

2.5.1. Dipole Moments

If a molecule has two large noninteracting dipoles and if their relative orientation varies as a function of torsional angle defining the conformers, the dipole moment can be a good probe to find the preferred conformation. The dipole moment of a molecule can usually be determined by measuring the dielectric constant of its solution dissolved in a nonpolar solvent.

The method is quantitative, or at least semiquantitative, in most cases. The accuracy of this method depends on the precision of the component bond moments and the accuracy of the estimated geometry of the assumed conformers. Since dipole moment is a vector quantity, the dipole moment of the mixture of conformers μ is described in terms of Eq. 2.11 in terms of mole fractions (p_i) and the dipole moments (μ_i) of component conformers.

$$\mu = (p_1\mu_1^2 + p_2\mu_2^2 + p_3\mu_3^2 + \cdots)^{1/2}$$

$$= \left\{ \sum p_i\mu_i^2 \right\}^{1/2} \tag{2.11}$$

The dipole moments were determined for a series of p-substituted 1-phenylethyl t-butyl sulfoxides, p-Y-C_6H_4CH(CH_3)SOBu^t.[51] Table 2.10 lists the data, together with the calculated dipole moments, on the basis of the geometries from MM and bond moment values in Table 2.11.

The dipole moment of the (RR)/(SS)-sulfoxide increased on replacement of the substituent Y from H to Br and then to NO_2. This is reasonable, because the group moment of Y increases in this order. The substituent effect for the (RS)/(SR)-diastereomers is anomalous in that the dipole moment decreases at first and then increases. The behavior is rationalized by assuming that the angle between S–O and C_{arom}–Y dipoles is greater than the right angle for (RS)/(SR)-sulfoxide (Fig. 2.18). This conclusion was reached by other spectral means and crystallographic determinations [O/Ph torsional angle 50° and 168°, respectively, for p-Br-(RR)/(SS)- and (RS)/(SR)-sulfoxides].

Dipole moments are useful when the orientation of the component bond (or group) moments are fixed. Thus, the method was applied to determine the predominant conformers of benzyl and 1-phenylethyl ketones. The preferred conformations of these ketones were also tabulated in Table 2.10. As to 1-phenylethyl ketone, the Ph/Bu^t gauche conformer (**A′**), favorable for CH/π interaction, was shown to be predominant. Benzyl t-butyl ketone seemed to exist as an equilibrium mixture of the two conformers **D′** and **E′**. The ap-conformer which is unable to have CH/π interaction was shown to predominate over the sc-conformer in this case. However, the energy difference

TABLE 2.10. Dipole Moments of Benyzl *t*-Butyl Sulfoxides and Benzyl *t*-Butyl Ketones *p*-XC$_6$H$_4$CHRZC(CH$_3$)$_3$

Compounds				Dipole Moment (D)		Estimated Conformer[a]		
X	R	Z	Configuration	Obsd.	Calcd.	Dipole	X-ray	MM
				Sulfoxides				
H	H	SO		3.90	3.85	E		D
Br	H	SO		4.72	4.61	E	E	
NO$_2$	H	SO		6.44	6.62	E		
H	Me	SO	(*RR/SS*)	3.84	3.75	A		A
Br	Me	SO	(*RR/SS*)	4.23	4.23	A	A	
NO$_2$	Me	SO	(*RR/SS*)	5.69	5.91	A		
H	Me	SO	(*RS/SR*)	3.86	3.28	A + B, C		A
Br	Me	SO	(*RS/SR*)	3.42	3.31	A + B, C	A	
NO$_2$	Me	SO	(*RS/SR*)	4.26	3.29	A + B, C		
				Ketones				
H	H	C=O		2.57	2.81	E' + D'		E'
Cl	H	C=O		3.68	3.82	F + D		
H	Me	C=O		2.49	2.51	A'		A'
Cl	Me	C=O		2.55	2.53	A'		

[a] Conformers: (**A**), (**B**), and (**C**) for (*RR/SS*) isomer of the sulfoxide are shown. For (*RS/SR*) isomer, O should attach at the asterisked positions (∗).

(**A**)   (**B**)   (**C**)   (**D**)   (**E**)

(**A'**)   (**D'**)   (**E'**)

TABLE 2.11. Bond Moment Values (m) Used for the Calculations

Bond or Group (+)-(−)	m/D	Bond or Group (+)-(−)	m/D
H–C	0.0	C–S	0.90
C–C$_6$H$_5$	0.35	C–C$_6$H$_4$Br(p)	1.53
C–C$_6$H$_4$NO$_2$(p)	4.00	S–O (sulfoxide)	3.00

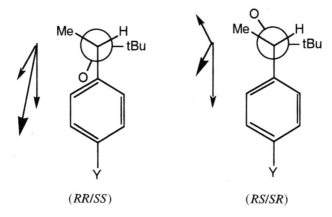

(*RR/SS*)   (*RS/SR*)

Figure 2.18. Conformations suggested for sulfoxide diastereomers by dipole moment measurements.

between the two conformers is relatively small. Assisted by the favorable entropic contribution, a considerable amount of the sp-conformer was shown to exist by IR spectroscopy.

2.5.2. Formation Constants

A chemical reaction should be favored when the product or the transition state is stabilized by this type of interaction.[52] Rate data and formation constants[53] may therefore provide evidence for the presence of the CH/π interaction.

2.5.2.1. Substrate Specificity of Calix[4]resorcarenes. Aoyama et al. studied the specificity of calix[4]resorcarene derivatives **21a–21c** with a series of ammonium salts and t-butanol (Table 2.12).[54]

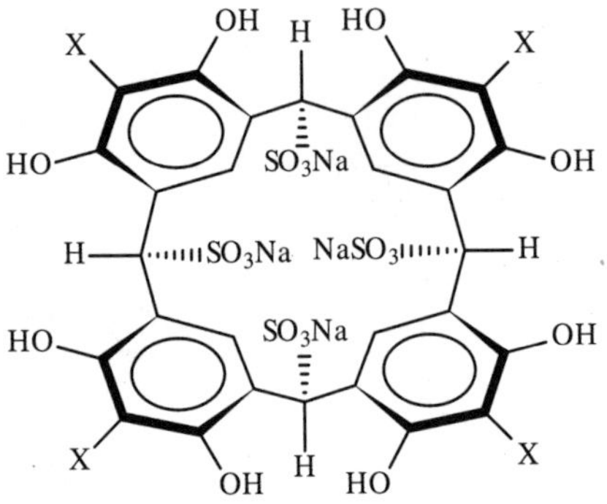

21

TABLE 2.12. Binding Constants (K mol^{-1}) for the Complexation of 21 with Various Guests in D$_2$O at 25°C

Guest	N^+H_4	MeN^+H_3	$Me_2N^+H_2$	Me_3N^+H	Me_4N^+	Me_3COH
21a (X=H)	1	1	3	30	160	4
21b (X=CH$_3$)					1500	19
21c (X=OH)					1800	24

The stability of the complex has been shown to increase with progressive methylation of the ammonium chloride guest. This cannot be a consequence of a mere bulk or simple electrostatic effect, since the replacement of X in the host from H **21a** to a more electron-donating group, such as CH$_3$ **21b** or OH **21c**, resulted in remarkable increases in the stability of the complexes. This is consistent with the expectation that orbital interaction becomes more prominent if the π-electron density of the aromatic ring increases. Trimethyl ammonium ion (Me$_3$N$^+$H, $K = 30$) was shown to be more specific than t-butyl alcohol (Me$_3$COH, $K = 4$) as a ligand. This is understandable because the complexing ability of the guest will increase if the hydrogens in CH$_3$ become more electron deficient (Fig. 2.19, X = C or N$^+$).

2.5.3. Vapor Pressure Measurement

Similar to hydrogen bonding, CH/π interaction will stabilize components easier in solution than in their pure state. This causes evolution of heat upon mixing. Evolution of heat lowers the energy of solution

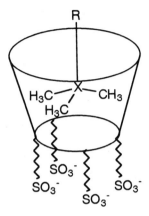

Figure 2.19. Complexation of **21** with Me_3XR ($X = N^+$ or C).

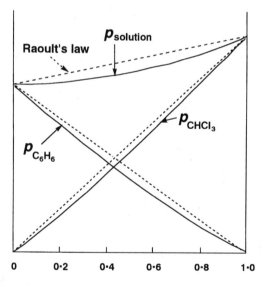

Figure 2.20. Schematic illustration of C_6H_6–$CHCl_3$ binary mixture.

and, as a result, the vapor pressure of components becomes lower than that predicted by Raoult's law. An example is given in Figure 2.20. Depression of vapor pressure in comparison to the ideal solution attests to the operation of an attractive CH/π interaction between chloroform and benzene.

2.5.4. Other Classical Methods

Tamres reported that an aromatic compound which gave out large heat on mixing with chloroform also showed a large shift in frequency of the O–D stretching band of deuterated methanol (CH_3OD).[55] Since the shift of the O–D stretching mode is related to hydrogen bonding, this can be considered as evidence that the heat of mixing is caused by the formation of a hydrogen bond. They also found the association of aromatic compounds with chloroform to occur in a 1 : 1 ratio.

REFERENCES

1. L. A. Woodward, *Introduction to the Theory of Molecular Vibrations and Vibrational Spectroscopy*, Oxford University Press, Oxford, UK, 1972; G. Herzberg, *Infrared and Raman Spectra of Polyatomic Molecules*, Van Nostrand, New York, 1945.

2. J. W. Emsley, J. Feeney, and L. H. Sutcliffe, *High Resolution Nuclear Magnetic Resonance Spectroscopy*, Pergamon Press, Oxford, 1966.

3. C. Djerassi, *Optical Rotatory Dispersion*, McGraw-Hill, New York, 1960.

4. C. P. Smyth, *Dielectric Behavior and Structure*, McGraw-Hill, New York, 1955.

5. P. J. Breen, J. A. Warren, E. R. Bernstein, and J. I. Seeman, *J. Am. Chem. Soc.*, **109**, 3453 (1987); J. A. Dickinson, P. W. Joireman, R. T. Kroemer, E. G. Robertson, and P. Simons, *J. Chem. Soc., Faraday Trans.*, 1467 (1997).

6. N. Sheppard, *Hydrogen Bonding*, D. Hadzi and H. W. Thompson, Eds., Pergamon Press, London, 1959, pp. 85–105.

7. L. J. Bellamy, *Infra-red Spectra of Complex Molecules*, 3rd ed., Chapman & Hall, London, 1975, pp. 13–104.

8. S. Pinchas, *Anal. Chem.*, **27**, 2 (1955); *ibid.*, **29**, 334 (1957); *Chem. & Ind.*, 1451 (1959); *J. Phys. Chem.*, **67**, 1862 (1963); H. Satonaka, K. Abe, and M. Hirota, *Bull. Chem. Soc. Jpn.*, **61**, 2031 (1988); J. L. Adcock and H. Zhang, *J. Org. Chem.*, **60**, 1999 (1995).

9. A. Allerhand and P. von R. Schleyer, *J. Am. Chem. Soc.*, **85**, 1715 (1963).

10. M. L. Josien and G. Sourisseau, *Bull. Soc. Chim. Fr.*, 178 (1955); M. L. Josien, G. Sourisseau, and C. Castinel, *ibid.*, 1539 (1955); N. Fuson, P. Pineeau, and M. L. Josien, *ibid.*, 423 (1957); M. L. Josien and G. Sourisseau, *Hydrogen Bonding*, D. Hadzi and H. W. Thompson, Eds., Pergamon Press, London, 1959, pp. 129–137.

11. R. West and C. S. Kraihanzel, *J. Am. Chem. Soc.*, **83**, 765 (1961).

12. J. C. D. Brand, C. Eglinton, and J. F. Norman, *J. Chem. Soc.*, 2526 (1961).

13. C. M. Huggins and G. C. Pimentel, *J. Phys. Chem.*, **60**, 1615 (1956).

14. D. Kivelson, S. Winstein, P. Bruck, and R. L. Hansen, *J. Am. Chem. Soc.*, **83**, 2938 (1961).

15. M. Karatsu, H. Suezawa, K. Abe, M. Hirota, and M. Nishio, *Bull. Chem. Soc. Jpn.*, **59**, 3529 (1986).

16. M. Hirota, T. Sekiya, K. Abe, H. Tashiro, M. Karatsu, M. Nishio, and E. Osawa, *Tetrahedron*, **39**, 3091 (1983).

17. Y. Nakai, G. Yamamoto, and M. Ōki, *Chem. Lett.*, 92 (1987); Y. Nakai, K. Inoue, G. Yamamoto, and M. Ōki, *Bull. Chem. Soc. Jpn.*, **62**, 2923 (1989).

18. L. W. Reeves and W. G. Schneider, *Can. J. Chem.*, **35**, 251 (1957).

19. R. F. Zurcher, *Helv. Chim. Acta*, **44**, 1380 (1961); *ibid.*, **46**, 2054 (1963); N. S. Bhacca and D. H. Williams, *Tetrahedron Lett.*, 3127 (1964); D. H. Williams and N. S. Bhacca, *Tetrahedron*, **21**, 2021 (1965); C. E. Johnson, Jr. and F. A. Bovey, *J. Chem. Phys.*, **29**, 1012 (1958).

20. N. Nakagawa and S. Fujiwara, *Bull. Chem. Soc. Jpn.*, **33**, 1634 (1960); R. E. Richards and J. V. Hatton, *Trans. Faraday Soc.*, **57**, 28 (1961).

21. N. Nakagawa, *Nippon Kagaku Zasshi*, **82**, 141 (1961); N. Nakagawa and S. Fujiwara, *Bull. Chem. Soc. Jpn.*, **34**, 143 (1961). See also M. Nishio, *Chem. Pharm. Bull.*, **17**, 262 (1969).

22. R. J. Abraham, *Mol. Phys.*, **4**, 369 (1961); P. Laszlo, *Progr. Nucl. Magn. Reson.*, **3**, 348, 376 (1967); C. J. Creswell and A. L. Allred, *J. Phys. Chem.*, **66**, 1469 (1962).

23. R. Ehama, A. Yokoo, M. Tsushima, T. Yuzuri, H. Suezawa, and M. Hirota, *Bull. Chem. Soc. Jpn.*, **66**, 814 (1993).

24. C. C. Hinckley, *J. Am. Chem. Soc.*, **91**, 5160 (1969); *J. Org. Chem.*, **35**, 2834 (1970); J. K. M. Saunders, S. W. Hanson, and D. H. Williams, *J. Am. Chem. Soc.*, **94**, 5325 (1972).

25. H. M. McConnell and R. E. Robertson, *J. Chem. Phys.*, **29**, 1361 (1958).

26. Perturbation to the conformational equilibria occurs in the cases of compounds bearing alkyl groups on both termini of the molecule, thus illustrating the importance of the CH/π interaction.

27. M. Nishio and M. Hirota, *Tetrahedron*, **45**, 7201 (1989).

28. Y. Kodama, K. Nishihata, and M. Nishio, *J. Chem. Res. (S)*, 102 (1977).

29. M. R. Willcott, III, R. E. Lenkinski, and R. E. Davis, *J. Am. Chem. Soc.*, **94**, 1744, 1779 (1972).

30. To avoid confusion, the conventional *threo/erythro* notation is used. The nomenclature according to the sequence rule does not correlate with the stereochemically corresponding structures.

31. Y. Kodama, S. Zushi, K. Nishihata, M. Nishio, and J. Uzawa, *J. Chem. Soc., Perkin 2*, 1306 (1980).

32. Y. Kodama, K. Nishihata, S. Zushi, M. Nishio, J. Uzawa, K. Sakamoto, and H. Iwamura, *Bull. Chem. Soc. Jpn.*, **52**, 2661 (1979); J. Uzawa, S. Zushi, Y. Kodama, Y. Fukuda, K. Nishihata, K. Umemura, M. Nishio, and M. Hirota, *ibid.*, **53**, 3623 (1980).

33. An exception is the *threo* alcohol bearing methyl as R; rotamer **B**, in which R is flanked by both benzylic methyl and the phenyl group, is the most stable. The possibility that the complexation with LSR perturbs the rotameric equilibria of the alcohols was excluded by monitoring the vicinal coupling constant; $^3J_{HH}$, remained unchanged by the addition of the lanthanide species.

34. J. Sicher, M. Chérest, Y. Gault, and H. Felkin, *Collect. Czeck. Chem. Commun.*, **28**, 72 (1963); S. Zushi, Y. Kodama, Y. Fukuda, Y. Takeuchi, M. Nishio, and H. Felkin, unpublished results.

35. J. H. Noggle and R. E. Schirmer, *The Nuclear Overhauser Effect, Chemical Applications*, Academic Press, New York, 1971.

36. H. Suezawa, A. Mori, M. Sato, R. Ehama, I. Akai, K. Sakakibara, M. Hirota, M. Nishio, and Y. Kodama, *J. Phys. Org. Chem.*, **6**, 399 (1993); H. Suezawa, R. Ehama, T. Hashimoto, and M. Hirota, unpublished data.

37. The NOE enhancement is denoted as $f H_{obs}(H_{irr})$, where H_{obs} and H_{irr} are the observed and the irradiated protons in the NOE experiment.

38. The ring size of an intramolecular CH/π chelate is counted by assuming that a weak hydrogen bond is present between the hydrogen atom of donor CH group and the *ipso*-carbon atom (C_1 of the aromatic ring).

39. V. L. Rosenfeld, *Z. Phys.*, **52**, 161 (1928).

40. M. Fetizon, I. Hanna, A. I. Scott, A. D. Wrixon, and T. K. Devon, *Chem. Commun.*, 545 (1971).

41. S. Araki, K. Sakakibara, M. Hirota, M. Nishio, S. Tsuzuki, and K. Tanabe, *Tetrahedron Lett.*, 6587 (1991); S. Araki, T. Seki, K. Sakakibara, M. Hirota, M. Nishio, and Y. Kodama, *Tetrahedron: Asym.*, **4**, 555 (1993).

42. J. Hudec and D. N. Kirk, *Tetrahedron*, **32**, 2475 (1976).

43. Y. Iitaka, Y. Kodama, K. Nishihata, and M. Nishio, *Chem. Commun.*, 384 (1974); Y. Kodama, K. Nishihata, M. Nishio, and Y. Iitaka, *J. Chem. Soc., Perkin 2*, 1490 (1976).

44. I. L. Karle, *Acta Cryst. B*, **28**, 2000 (1972).

45. Numbers reported here were obtained, by program CHPI (Chapter 11), with the use of coordinates as they appeared in the literature.

46. A. J. deKok and C. Romers, *Acta Cryst. B*, **30**, 1695 (1974).

47. A. J. deKok and C. Romers, *Acta Cryst. B*, **31**, 1535 (1975).

48. A. J. deKok, C. Romers, and J. Hoogendorp, *Acta Cryst. B*, **30**, 2818 (1974).

49. F. H. Allen, J. E. Davies, J. J. Galloy, O. Johnson, O. Kennard, C. F. Macrae, E. M. Mitchell, G. F. Mitchell, J. M. Smith, and D. G. Watson, *J. Chem. Inf. Comput. Sci.*, **31**, 187 (1990); G. Ferraris, D. W. Jones, J. Yerkess, and K. D. Bartle, *J. Chem. Soc.*, *Perkin 2*, 1628 (1972). Y. Umezawa, S. Tsuboyama, K. Honda, J. Uzawa, and M. Nishio, *Bull. Chem. Soc. Jpn.*, **71** (1998) in press.

50. E. L. Eliel, N. L. Allinger, S. J. Angyal, and G. A. Morrison, *Conformational Analysis*, Interscience, New York, 1965.

51. M. Hirota, Y. Takahashi, M. Nishio, and K. Nishihata, *Bull. Chem. Soc. Jpn.*, **51**, 2358 (1978).

52. Review: H. Okawa and S. Kida, *Kagaku no Ryoiki*, **37**, 276 (1983); H. Okawa, *Coord. Chem. Rev.*, **92**, 1 (1988).

53. M. Vincenti, E. Dalcanale, P. Soncini, and G. Guglielmetti, *J. Am. Chem. Soc.*, **112**, 445 (1990); M. Vincenti, E. Pelizetti, E. Dalcanale, and P. Soncini, *Pure & Appl. Chem.*, **65**, 1507 (1993); A. Arduini, M. Cantoni, E. Graviani, A. Pochini, A. Secchi, A. R. Sicuri, R. Ungaro, and M. Vincenti, *Tetrahedron*, **51**, 599 (1995).

54. K. Kobayashi, Y. Asakawa, and Y. Aoyama, *Supramolec. Chem.*, **2**, 133 (1993).

55. M. Tamres, *J. Am. Chem. Soc.*, **74**, 3375 (1952).

CHAPTER 3

QUANTUM-MECHANICAL TREATMENT OF CH/π INTERACTION

3.1. INTRODUCTION

The CH/π interaction is one of the weakest extremes of the so-called "weak molecular interactions."[1,2] Weakly interacting systems have been treated by molecular orbital calculations with various degrees of approximation. The quantum chemical treatment can be classified into two categories: the variational and the perturbational methods.[3]

3.1.1. Theoretical Treatments of Weak Molecular Interactions

In the variation method, a whole interacting system (R $\cdots$ S) is treated as a supermolecular system and solved by using various versions of molecular orbital theory. In the case of intramolecular interaction, the variational method is usually adopted. The interaction energy (ΔE) can be evaluated as the difference between the sum of the energies of the components (R and S) and the total energy (E_{total}) of the interacting system (R $\cdots$ S). The interaction energy can be formulated as follows:

$$\Delta E = E_{total} - (E_R + E_S) \tag{3.1}$$

The interaction energy is generally a small difference between very large energies; nevertheless, the ΔE values come out quite correctly as a result of compensation of errors between interacting and non-interacting systems. Thus, ab initio calculations[4,5] of various degrees of approximation have given reliable interaction energies in most cases.

The CH/π interaction is often assumed to be very weak hydrogen bonding, in which the contribution of the dispersion force plays an important role. In such a situation, the electron correlation must be included, which deters reliable calculations of the large systems. Features of CH/π and other weak intramolecular interactions obtained by unsophisticated types of calculations should hence be regarded as of qualitative value. Some examples are given in Section 3.3.

The second method for the theoretical treatment of a weakly interacting system is based on the perturbation theory. A modification of the perturbation calculation, as proposed by Morokuma et al.[6,7] and later by Kollman,[8] has been widely applied as an alternative practical method to analyze such interacting systems. This technique evaluates the interaction energy as the sum of (1) electrostatic (Coulombic) interaction E_C, (2) polarization (induction E_I and dispersion E_D), (3) exchange repulsion E_{ER}, (4) charge-transfer interaction (delocalization) E_{CT}, and (5) cross terms, which allows us to discuss the origin of interaction quantitatively:

$$\Delta E = E_C + E_I + E_{ER} + E_D + E_{CT} + E(\text{cross terms}) \qquad (3.2)$$

3.1.2. Evaluation of Component Interaction Energy Terms

3.1.2.1. Coulombic Interactions. Hydrogen-bond-like weak interactions are originated to a large extent from electrostatic terms, which comprise a long-range force. Distribution of charge in a molecule can be reproduced by the positive-point charges at the positions of nuclei and the negative charge of electrons described by the wave functions.

The E_C term arises from the Coulombic interaction within the static charge distribution, which can be approximated by the multipole expansion. In effect, the interaction energy can be expanded into interaction terms between point charges and multipoles; the E_C term involves the charge–charge (r^{-1} term), charge–dipole (r^{-2} term), charge–quadrupole (r^{-3} term), dipole–dipole (r^{-3} term), dipole–quadrupole (r^{-4} term) and so on:

$$E_C = q_R q_S/r + (1/r^2)[q_R \mu_S \cos \theta_S + q_S \mu_R \cos \theta_R] + \cdots \qquad (3.3)$$

In the case of CH/π interaction, dipole–quadrupole interaction can be an important term among various Coulombic terms. Assuming

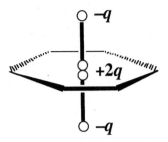

Figure 3.1. Bar quadrupole in the benzene molecule.

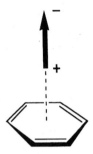

Figure 3.2. The most stable orientation of the benzene quadrupole relative to the dipole [C($-$)-H($+$)].

that the charge distribution in benzene is a bar-quadrupole, as illustrated in Figure 3.1, Nakagawa et al.[9] concluded that the most stable orientation of the benzene ring relative to the dipole (shown by the arrow) is T-shaped, as shown in Figure 3.2. The dipole–quadrupole interaction was estimated to be about 14 times as large as the interaction between the permanent dipole (C–H) and the induced dipole (C$_6$H$_6$) of the same system when CH and C$_6$H$_6$ are located at their closest distance.

3.1.2.2. Inductive and Dispersion Interactions.
In addition to the static Coulombic interaction, the electric charges in the approaching molecules (R and S) can cause a mutual dynamic deformation of their electron distributions, inducing an extra interaction between the molecules. This sort of interaction produces an attractive force which is proportional to the polarizabilities of the interacting molecules.

The interaction between charge-induced dipole and charge is called inductive interaction and can be expressed by Eq. 3.4:

$$E_I = -(1/2)\alpha_R\{q_S^2/r^4 + \mu_S^2[3\cos^2\theta_S + 1]/r^6 + \cdots\}$$
$$-(1/2)\alpha_S\{q_R^2/r^4 + \mu_R^2[3\cos^2\theta_R + 1]/r^6 + \cdots\} \quad (3.4)$$

Similarly, the interaction between the two induced dipoles (multipoles), E_D, is expressed by Eq. 3.5:

$$E_D = -C_6/r^6 - C_8/r^8 - C_{10}/r^{10} - \cdots \qquad (3.5)$$

where C_6, C_8, and C_{10} are appropriate constants, and all terms produce additional attractive forces; E_D is called dispersion energy and corresponds to the attractive term of the well-studied van der Waals interaction, which is dominated by the r^{-6} term. Hence, E_D is expected to be a short-range interaction. The van der Waals interaction potential has been described by Lennard-Jones empirically in terms of distance between nonbonded particles (Eq. 3.6):

$$E_{vdw} = -a/r^6 + b/r^n \quad (n = 9\text{--}12) \qquad (3.6)$$

The r^{-6} term was related to the polarizabilities (α_R and α_S) and the ionization potentials (I_R and I_S) of the interacting systems R and S by London:[10]

$$E_D = -(3/2)(\alpha_R \alpha_S/r^6)[I_R I_S/(I_R + I_S)] \qquad (3.7)$$

It is quite laborious and time-consuming to calculate dispersion interactions accurately by molecular orbital theory involving electron correlation. However, they can be estimated by a simple second-order perturbation treatment given by Eq. 3.8:[11]

$$E_D = -4\sum_{i(R)}^{occ} \sum_{j(R)}^{unoc} \sum_{k(S)}^{occ} \sum_{l(S)}^{unoc} (ij/kl)^2/[E_{ij}(R) - E_{kl}(S)] \qquad (3.8)$$

3.1.2.3. Delocalization or Charge Transfer Interaction.

Delocalization interaction is assumed to operate over a relatively short range and plays some role in weak interactions like CH/π if the distance between the interacting H and C_{sp^2} atoms is less than the sum of van der Waals radii and if the geometrical arrangements of the CH and π-system is suitable. This interaction has often been treated as a perturbation between the interacting systems (R and S) rather than by the variational method.[12] The perturbation calculation is based on the independent MO calculations on R and S, thus, is far less time-consuming than the full calculation on the whole system. If we assume that the interaction occurs between the atom r of molecule R and the atom s of molecule S, the delocalization energy due to this interaction can be evaluated by Eq. 3.9:

$$E_{deloc} = \sum_k^{occ} \sum_n^{unoc} (C_{rn} C_{sk} \Delta \beta_{rs})^2 / (E_k - E_n)$$

(CT from S to R)

$$+ \sum_m^{occ} \sum_l^{unoc} (C_{rm} C_{sl} \Delta \beta_{rs})^2 / (E_m - E_l)$$

(CT from R to S)   (3.9)

where C_{rm} and C_{rn} refer to the coefficients of occupied and unoccupied MOs of R and C_{sk} and C_{sl} refer to those of occupied and unoccupied MOs of S, respectively. If we assume that R and S are XH donor and π-acceptor, respectively, the first term of Eq. 3.9 becomes the largest and the charge-transfer from S to R becomes the most important. Thus, in a simple perturbation model for CH/π interaction, only the interaction between the highest occupied π-orbital and antibonding (σ^*) orbital of X–H group is taken into account and the interaction energy is given by Eq. 3.10:

$$E_{deloc} = (S_{rs} \Delta \beta)^2 / (E_\pi - E_{CH^*}) \qquad (3.10)$$

This implies that the charge-transfer (partial electron transfer) from the π-system to XH occurs during the process of interaction, and a very weak covalent bond is formed. Since the interaction energy increases as the energy difference $\Delta E (= E_\pi - E_{XH^*})$ decreases, both the fall of E_{XH^*} caused by introducing an electronegative group on the X atom and the rise of E_π caused by introducing an electron-donating group on the π-system will favor the interaction.

3.1.2.4. Similarity with Hydrogen Bonds.
Just like hydrogen bonding, CH/π interaction is defined phenomenologically to give a general rationalization to the widely observed affinity of C–H hydrogen atoms and π-electron systems. To gain a deeper understanding of the nature of CH/π interaction, both interactions are characterized quantum chemically by comparing the above-mentioned energy terms.

TABLE 3.1. Energy Terms Contributing to Hydrogen Bonding and Relevant Interactions[a]

XH/Y Interaction	OH/O[b]	NH/π[c]	CH/O[b]	CH/π[d]
Coulombic (ES)	− 10.5	− 2.5	− 0.5	− 0.14
Delocalization (CT)	− 2.4		− 0.9	− 0.72
Polarization (PL)	− 0.6		− 0.1	− 0.02
Exchange repulsion (ER)	6.2		0.5	0.24
Total (H-bond energy)	− 7.8	− 3.5	− 1.1	− 0.88

[a] Energy terms are given in kcal mol^{-1}.
[b] K. Morokuma, *Acc. Chem. Res.*, **10**, 294 (1977).
[c] M. F. Perutz, *Phil. Trans. Roy. Soc.*, *A*, **345**, 105 (1993).
[d] T. Takagi, A. Tanaka, S. Matsuo, H. Maezaki, M. Tani, H. Fujiwara, and Y. Sasaki, *J. Chem. Soc.*, *Perkin 2*, 1015 (1987).

In a XH $\cdots$ Y hydrogen bond, the attractive force that holds the donor H and acceptor Y atoms in an unusually short distance is known to originate from a cooperative effect of the Coulombic, the delocalization, and unusually small repulsion forces. The magnitudes of these terms were evaluated with the intermolecular hydrogen bonding of water by Coulson,[13] and more recently by Morokuma (as given in Table 3.1). In the case of OH/O, a relatively small repulsion term (ER) allows the approach of (O) H and O atoms far within the sum of their van der Waals radii.

The electrostatic stabilization energy (ES) decreases abruptly in the order OH/O > CH/O > CH/π. In contrast, the delocalization energy (CT) decreases only moderately in this order, and becomes most important in the CH/π interaction. The figures in Table 3.1 show that the CT term in CH_4/C_2H_4 is approximately one-third as large as in a very strong hydrogen bond. Because the (C)H $\cdots$ X distance in the CH/O and CH/π interacted systems are not significantly shorter than the sum of their van der Waals radii, the ER term does not contribute seriously.

3.1.3. Geometrical Preference of CH/π Interaction

The geometrical arrangement of the interacting CH and π components should be important since the overlap integral occurs in

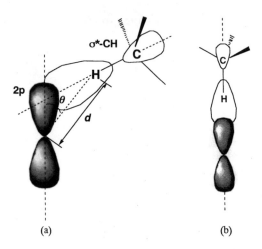

Figure 3.3. Geometrical arrangement of interacting X–H bond and 2p-orbital. (a) General noncoaxial approach and (b) coaxial approach (dark and white lobes denote occupied and unoccupied orbitals, respectively).

Eq. 3.10. The vacant antibonding C–H orbital stretches outside the hydrogen atom along the C–H axis. The counterpart π-orbital (essentially the $2p_z$-orbital of carbon) has a well-known dumbbell shape and stretches perpendicular to the π-plane. The magnitude of an overlap integral should be a function of both distance and orientation of these two orbitals.[14]

In the simple case, the integral can be expressed by the angle of the intersecting axes of XH and p-orbitals (θ) and the distance (d) in Figure 3.3a. When the XH group approaches coaxially from the direction perpendicular to the π-molecular plane (as in Fig. 3.3b), the stabilization energy becomes maximum. Examples of CH/π interacted systems are given in Section 3.2.

3.2. MOLECULAR ORBITAL CALCULATIONS OF CH/π AND OTHER WEAK INTERACTIONS

3.2.1. Intermolecular CH/π Interactions

The concept of CH hydrogen bonding between an acidic C–H group and an aromatic compound was first introduced in order to interpret the aromatic solvent-induced shift of the ^{1}H-NMR spectra of certain organic compounds bearing acidic CH group.[15] Ever since such compounds as chloroform and other tri(negative atom)-substituted methanes have been shown to interact with π-bases, a number of model calculations have been carried out on binary systems consisting of CH acid and π-bases with the aim of characterizing the interaction involved. The calculations usually involved chloroform, hydrogen cyanide, acetylene as the acidic C–H component, and ethylene, acetylene, benzene as the π-component.

3.2.1.1. Electrostatic Interaction Models.

As in other discussions on weak hydrogen bonds, the conclusions on the nature of the CH/π interaction are somewhat controversial. Some focus on the importance of electrostatic force and others stress the participation of delocalization force. Price and Stone[16] evaluated the minima of electrostatic energies for benzene–acetylene, s-tetrazine–acetylene, "benzene dimer," and several other bimolecular systems by using sets of distributed multipoles obtained from ab initio wave functions of the components (the Buckingham–Fowler model[17]). In the benzene–acetylene system (Table 3.2), the geometry in which the acetylenic CH is directed to a C–C bond of benzene is shown to be the most stable, giving a stabilization energy as high as $-2.1\,\mathrm{kcal\,mol^{-1}}$. Similar calculations on the s-tetrazine–acetylene system showed that the CH/N (lone pair) hydrogen-bonded geometry is the most stable. A T-shaped geometry is given to the benzene dimer **1**.[18] The T-shaped acetylene dimer **2** was also shown to be the most stable by ab initio calculation.[19]

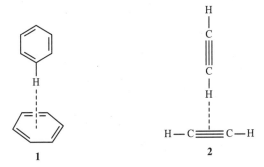

1   **2**

TABLE 3.2. Minima of the Buckingham–Fowler Potential for the Acetylene–Benzene System[a]

Geometry	Distance (Å)	Energy (kcal mol^{-1})
H-bonded to C–C of ring (I) (perpendicular)	4.1	−2.1
H-bonded to triple bond (II) (perpendicular)	4.9	−1.3
H-bonded to triple bond (III) (coplanar)	4.9	−0.8
Edge to edge (IV) (perpendicular)	4.7	−1.2

[a] Geometries are shown in Figure 3.4.

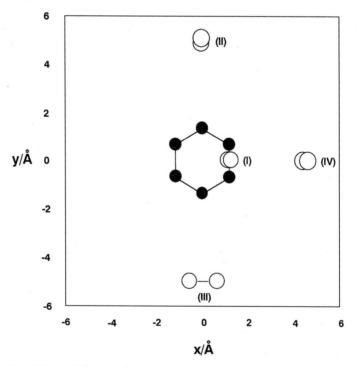

Figure 3.4. The minimum electrostatic energy structures for the benzene–acetylene van der Waals complex. Structures I–IV correspond to the structures in Table 3.2. (In structure I, the acetylene molecule is located 4.1 Å above and perpendicular to a C–C bond of the benzene molecule.)

3.2.1.2. Charge-Transfer Interaction Models. On the other hand, the CH/π complex has been assumed to be a weak charge-transfer complex. Nakagawa et al. reported the interaction energy for the most stable methane–benzene system to be -0.8 kcal mol^{-1} by CNDO/2 calculations.[20] Recently, more elaborate calculations were reported on CH_4–C_2H_2 (acetylene) **3** and CH_4–C_2H_4 (ethylene) binary systems by Takagi et al.[21] The report includes an ab initio calculation using a 4-31 G basis set on the CH_4–C_2H_4 supermolecular system **4** in which one of the C–H bonds of methane is fixed perpendicular to the molecular plane of C_2H_4 and directed to the center of the C=C bond, while optimizing the intermolecular distance R_{CC}. The energy minimum is located at $R_{CC} = 4.4$ Å in the case of CH_4–C_2H_4 (ethylene). The results were analyzed by employing Morokuma's variational method.[6] The large $CT_{B \to A}$ in comparison with the CH_4–C_2H_6 case implies that the charge-transfer from ethylene to methane contributes greatly to stabilizing the CH_4–C_2H_4 molecular system (Table 3.3), whereas the charge-transfer in the opposite direction ($CT_{A \to B}$) contributes little to the stabilization of the interacting system. The most stable geometry of CH_4–C_2H_2 and CH_4–C_2H_4 binary systems thus obtained are illustrated in Figure 3.5.

MP2-4/6-31G* or 6-311G** calculations were recently reported on acetylene-$CH_{4-n}Cl_n$ bimolecular systems[22] and the T-shaped geometry **3** was again shown to be the energy minimum (Table 3.4). The density-functional theory also gives energy minima of similar geometries except for the C_2H_2–CH_4 system.

TABLE 3.3. Calculated Energies (in kcal mol^{-1}) and Optimized Intermolecular Distances (R_{CC}) for the $CH_4(A)$–C_2H_n (B; $n = 2, 4, 6$) Systems^a

B	$R_{CC/nm}{}^b$	E_{int}	E_C	E_I	$CT_{A \to B}$	$CT_{B \to A}$	E_D
C_2H_2	0.44	-0.66	-0.13	-0.01	-0.02	-0.52	-0.19
$C_2H_4{}^c$	0.44	-0.88	-0.14	-0.02	0.00	-0.72	-0.24
C_2H_6	0.50	-0.24	-0.08	0.00	-0.08	-0.08	-0.19

^{a} E_{int}, stabilization energy in reference to the isolated A and B; E_c, Coulombic energy; E_I, polarization energy; $CT_{A \to B}$ and $CT_{B \to A}$, charge-transfer energy due to electron migration from A to B and B to A, respectively; E_D, dispersion energy.
^{b} Perpendicular distance between the carbon atom of A and the molecular axis of B (Fig. 3.5).
^{c} Calculation on geometry **4**.

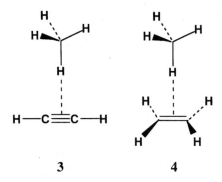

Figure 3.5. Energy minimum structures of CH_4–C_2H_2 (acetylene) and CH_4–C_2H_4 (ethylene) binary systems.

TABLE 3.4. Structural Parameters and Interaction Energies for Several C_2H_2/H-donor Systems

H-Donor	Calculation	CH $\cdots\pi$/Å	ΔE_{int} (kcal mol^{-1})
CH_4	MP2/6-31G*	3.080	− 0.10
	MP4/6-311G**	3.151	− 0.19
HCN	HF/6-311G**	2.826	− 1.64
	MP4/6-311G**	2.652	− 1.76
CH_3Cl	MP2/6-31G*	2.852	− 0.55
CH_2Cl_2	MP2/6-31G*	2.622	− 1.31
$CHCl_3$	MP2/6-31G*	2.495	− 1.82

3.2.2. Intramolecular CH/π Interactions

In molecular orbital calculations (1) the bond populations (bond orders) between nonbonded (CH/π interacted) atoms become a critical measure of the presence and the strength of an intramolecular interaction and (2) the total energy becomes a way to measure the stabilization of the system by the interaction.

For example, intramolecular OH/π interaction in norborneol **5** was verified by STO-3G calculations with full optimization of geometry.[23] The calculations showed that the syn-cis isomer capable of forming intramolecular OH/π hydrogen bond is the most stable. The attractive OH/π interaction was indicated by the elongated OH bond lengths (R_{OH}) and the shift of OH stretching frequency as given in Table 3.5.

TABLE 3.5. Energies and Some Geometric Parameters of Norborneol Isomers by Ab Initio Calculations

Isomer	syn-cis	syn-trans	anti-cis	anti-trans
	5a	5b	5c	5d
ΔE (kcal mol^{-1})	-1.68	$+1.56$	(0)	$+0.38$
R_{OH} (nm)	9.906	9.876	9.884	$+9.875$
ν_{OH} (cm^{-1})	3601	3639	3630	3639

The frequency shift of ν_{OH} compares well with the observed shift of the open-chain analog, viz. ν_{free} is 3635 cm^{-1} and ν_{int} is 3596 cm^{-1} for $CH_2{=}CHCH_2CH_2OH$.[24] The bond order between "nonbonded" OH $\cdots$ C=C in **5a** is as high as 1.8×10^{-2}, by far the largest among isomers **5b–5d**.

As for the intramolecular CH/π interaction, ab initio 4-31G//STO-3G calculations on the several conformers of 1-phenyl-2-propanol **6** showed that the most stable conformer is the OH/π interacted phenyl/methyl *anti* (and phenyl/OH *gauche*) geometry $C_{OH/\pi}$ (Table 3.6), and the OH/π interaction energy was estimated to be 2.2 kcal mol^{-1}. Experimentally, the presence of two different conformers was detected by IR spectroscopy. The energy difference between them was determined to be 1.0 kcal mol^{-1} from the temperature dependence of the intensities of the OH infrared absorption bands at 3626 cm^{-1} (free) and 3604 cm^{-1} (OH/π interacted). In the conformer $C_{OH/\pi}$, a small positive bond population (1.62×10^{-2}) was assigned to the bonding interaction between the OH proton and a ring carbon atom.

Similar positive-bond populations were observed between the methyl protons and aromatic carbon atoms of CH/π interacted phenyl/methyl *gauche* conformers A and B. The population analysis gave positive bond populations between nonbonded H in methyl and aromatic C atoms (0.48×10^{-2} and 0.74×10^{-2} for conformers A and B, respectively), about one half of the OH/π interaction. This suggests participation of the delocalization or charge-transfer force between CH_3 and the π-electrons of the aromatic ring in the CH/π-proximate conformers A and B.

TABLE 3.6. Relative Energies^a (ΔE; in reference to Me/Ph *ap*-conformer C) and Populations of Several Stable Conformations of 1-Phenyl-2-Propanol[25]

$$C_6H_5CH_2CH(CH_3)OH$$

6

Conformer			
Ph/OH	*sc*	*sc*	*ap*
Ph/CH$_3$	*sc*	*ap*	*sc*
	A	**C**^{b} **C**$_{OH/\pi}$	**B**
ΔE (kcal mol^{-1})	-0.64	0.00 -2.20	$+0.38$
Populations			
C(Ph)$_{mean}$ atom	6.167	6.159 6.171	6.161
OH/Ph bond ($\times 10^2$)	-0.03	-0.02 $+1.62$	-0.02
CH$_3$/Ph bond ($\times 10^2$)	$+0.48$	-0.03 $+0.03$	$+0.74$

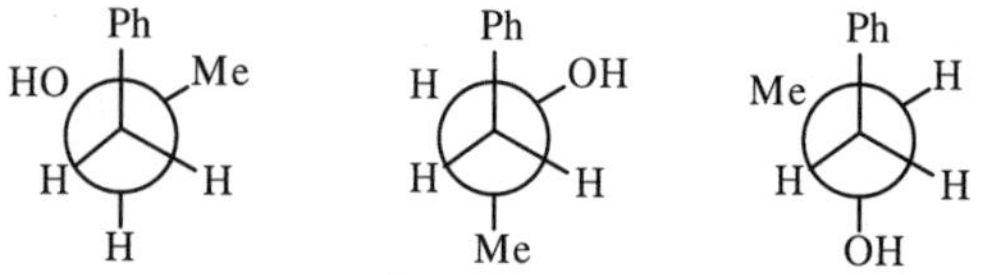

^{a} ΔE, in reference to Me/Ph *ap*-conformer C.
^{b} Hypothetical Ph/CH$_3$ *ap*-conformer without OH/π interaction.

α-Phellandrene, (*R*)-5-isopropyl-2-methyl-1,3-cyclohexadiene **7**,[26] is another example of an intramolecularly CH/π interacted system. MMP2 calculations[27] showed six energy minima (A–F in Fig. 3.6). Ab initio 4-31G//STO-3G calculations with full geometry-optimization were carried out starting from these geometries.

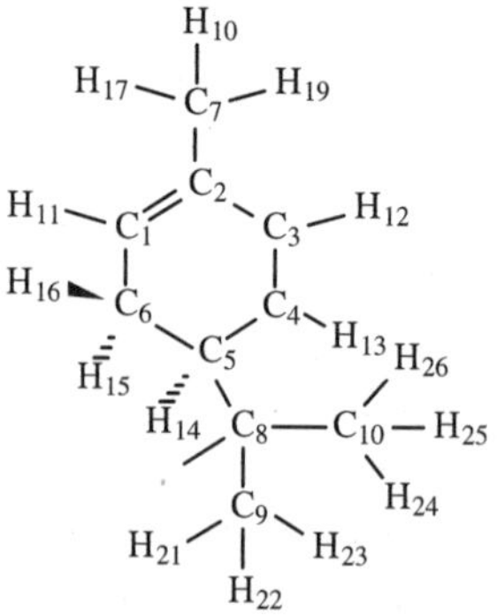

7

When a hydrogen atom of isopropyl group is located within the sum of the van der Waals radii from a sp^2 carbon, and the relative arrangements of the CH bond to the aromatic ring is appropriate (small θ in Fig. 3.3), CH/π interaction is expected to occur. The nonbonded C_{sp^2}–H pairs within 2.95 Å ($= 1.70$ of sp^2 carbon $+ 1.25$ of C_{sp^3}–H hydrogen) in the six energy minima are given in Table 3.7.

Nonzero and positive bond populations were observed between the proximate C/H pairs whenever θ is small. The C(4)/H(20) pair of conformer F has a negative bond population, because H(20) is

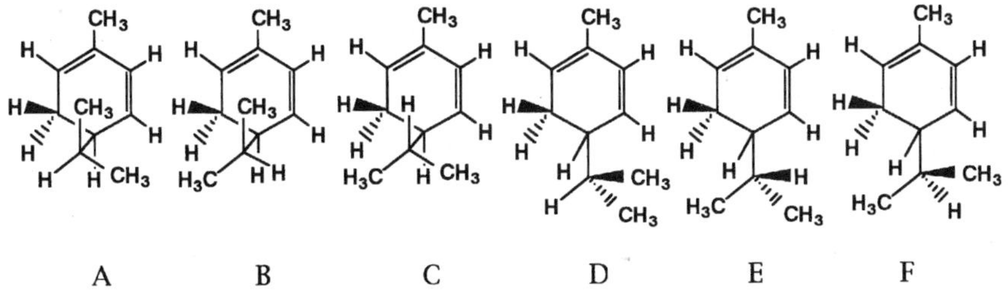

Figure 3.6. Six conformers of α-phellandrene 7.

TABLE 3.7. Overlap Populations (N_{CH}) Between Proximate Nonbonded $C_{sp^2 \cdots H}$ Pairs in the Six Stable Conformers of α-Phellandrene[a]

Conformer	C_{sp^2}-H Pair	N_{CH}	d(Å)	θ/(deg)
A	C(1)-H(23)	8.80×10^{-3}	2.80	17.5
	C(4)-H(22)	1.13×10^{-2}	2.86	21.1
	C(4)-H(26)	5.51×10^{-3}	2.85	43.8
B	C(3)-H(25)	5.18×10^{-3}	2.95	29.6
	C(4)-H(20)	-2.28×10^{-4}	2.62	49.8
	C(4)-H(25)	9.93×10^{-3}	2.81	23.5
C	C(1)-H(20)	1.18×10^{-2}	2.86	26.5
	C(4)-H(20)	7.11×10^{-3}	2.79	44.2
	C(4)-H(23)	1.18×10^{-2}	2.61	25.0
D	C(4)-H(25)	6.91×10^{-3}	2.77	43.2
E	C(4)-H(20)	-1.19×10^{-3}	2.73	54.3
	C(4)-H(26)	2.42×10^{-4}	2.79	83.6
F	C(4)-H(22)	7.58×10^{-3}	2.75	39.8
	C(4)-H(20)	-4.63×10^{-3}	2.72	87.0

[a] By 4.31G calculations.

located approximately on the nodal plane ($\theta = 87.0°$) of the p-orbital of C(4).

The nonbonded $C_{sp^2} \cdots H$ overlap population ($N_{C \cdots H}$) is sensitive to θ (Fig. 3.3), as shown in Figure 3.7. The plot suggests that the coaxial approach of the CH donor is the most favorable to the CH/π interaction.

The frontier molecular orbitals of the interacting system are suggestive as well: HOMO and LUMO of this molecule are essentially π-type contaminated with small amounts of the s-orbitals of hydrogen and saturated carbon atoms, and the mixing of the isopropyl

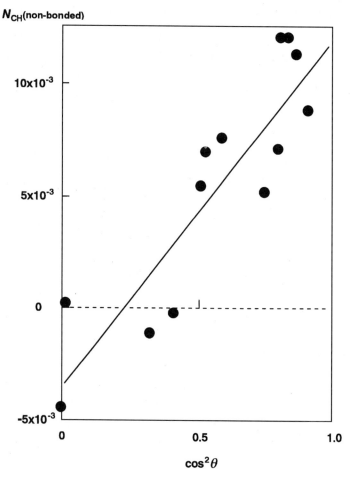

Figure 3.7. Angular dependency of nonbonded $C_{sp^2} \cdots H$ overlap populations in six stable conformers of α-phellandrene.

hydrogen orbitals is more visible whenever the molecule takes on a conformation capable of a CH/π interaction. The interaction was also shown to accompany some perturbation in the CH bonding electrons. In conformer A, the CH/π interacting C(10)–H(26) bond is slightly weaker than, for example, the C(10)–H(24) bond, which is void of such interaction, as shown by their bond populations, 0.769 versus 0.777. This result indicates that the CH/π bonding interaction is caused by partial charge-transfer from the π-donor to the C–H acceptor, or electronic excitation from occupied π-orbital to unoccupied antibonding C–H σ^* orbital.

3.3. SUBSTITUENT EFFECTS AS PROBES TO ASSESS THE NATURE OF CH/π INTERACTION

As the CH/π interaction is extremely weak, it is difficult to observe the perturbation of physical properties caused by this interaction. Therefore, IR, NMR and other spectroscopy, as well as X-ray crystallography, could not be applied in the same manner as the cases of typical hydrogen bonds. For example, infrared hydrogen bond shift of CH stretching absorption is very small even in the interaction of chloroform and π-bases. With intramolecular cases, reverse high frequency shift is often observed. In proton NMR studies, expected hydrogen bond shift is obscured by the strong high field shift due to the magnetic anisotropy induced by the aromatic ring.

In these circumstances, substituent effects on both CH and π counterparts should give crucial evidence for the weak bond formation via charge-transfer interaction. Electron-withdrawing substituents on the CH carbon lowers level of antibonding CH orbital and electron-donating substituents on the π system should raise the highest occupied π orbital; both the substituent effects favor the CH/π interaction by narrowing the energy gap of the interacting orbitals.

3.3.1. Electronic Substituent Effects on the Aromatic π-Systems

The electronic substituent effect on the equilibrium and the enthalpy of formation of weak bonding can be a good criterion to learn about the nature of interaction. As discussed in the previous section (3.2),

TABLE 3.8. Summary of the Substituent Effect on CH/π Interaction

Compound (System)	Method	Increasing Order of Feasibility of CH/π Interaction	References
Substituents on π-Bases, intramolecular			
$p\text{-XC}_6\text{H}_4\text{CHMeCOCDMe}_2$	IR(ν_{CD}), K_{eq}	X:NO_2 < Br < Cl < H ≈ Et < Me < NH_2	a
$p\text{-XC}_6\text{H}_4\text{CHMeCOCDMe}_2$	^{2}H-NMR	X:NO_2 ≈ Br < Cl < H ≈ Et < Me	a
$p\text{-XC}_6\text{H}_4\text{CH}_2\text{OCHMe}_2$	NOE	X:Br < H < MeO	b
$p\text{-XC}_6\text{H}_4\text{CH}_2\text{CONMeCHMe}_2$	NOE	X:NO_2 < H < MeO	b
$p\text{-XC}_6\text{H}_4\text{CH}_2\text{OCHO}$	NOE, LIS	X:NO_2 < Cl < H < Me < MeO	c
$p\text{-XC}_6\text{H}_4\text{CHMeOCHO}$	NOE, LIS	X:Cl < H < Me	c
X-Triptycene	NMR, δ_{CH_3}	X:NO_2 < H < NMe_2	e
Substituents on π-Bases, Intermolecular			
$\text{C}_6\text{H}_{6-n}\text{X}_n$/CHCl$_3$	NMR(δ_{CH}), K_{eq}	X_n:Cl < D < CD$_3$ < $p\text{-(CD}_3)_2$ < $o\text{-(CD}_3)_2$ < $(\text{CD}_3)_6$	d
$\text{C}_6\text{H}_{6-n}\text{X}_n$/CHCl$_3$	IR, $\Delta\nu_{CD}$	X_n: < H < CH_3 < $m\text{-(CH}_3)_2$ < $1,3,5\text{-(CH}_3)_3$ < $(\text{CH}_3)_4$ < $(\text{CH}_3)_6$	

Substituents on CH-Donor, Intramolecular

C$_6$H$_5$CH$_2$XCHMe$_2$	NOE	X:CO < NMe < O	[b]
C$_6$H$_5$CH$_2$XCHO	NOE	X:CH$_2$ < NMe < O	[b]
Y-Triptycene	NMR, δ_{CH_3}	Y:NO$_2$ < H < NMe$_2$	[e]

Substituents on CH-Donor, Intermolecular

C$_6$H$_6$/CHXYZ	NMR, δ_{CH}, K_{eq}	HC(OMe)$_3$ < HCHCl$_2$ < HCCl$_3$ < HCBr$_3$ < HCCOCCl$_3$	[b]
C$_6$H$_6$/CH$_3$C$_6$H$_4$Y	NMR, δ_{CH_3}	Y:NO$_2$ < Br < Cl < H < Me < NH$_2$	[f]

[a] M. Karatsu, H. Suezawa, K. Abe, M. Hirota, and M. Nishio, *Bull. Chem. Soc. Jpn.*, **59**, 3529 (1986).

[b] H. Suezawa, R. Ehama, T. Hashimoto, and M. Hirota, unpublished data.

[c] H. Suezawa, A. Mori, M. Sato, R. Ehama, I. Akai, K. Sakakibara, M. Hirota, M. Nishio, and Y. Kodama, *J. Phys. Org. Chem.*, **6**, 399 (1993).

[d] R. Ehama, A. Yokoo, M. Tsushima, T. Yuzuri, H. Suezawa, and M. Hirota, *Bull. Chem. Soc. Jpn.*, **66**, 814 (1993).

[e] Y. Nakai, K. Inoue, G. Yamamoto, and M. Oki, *Bull. Chem. Soc. Jpn.*, **62**, 2923 (1989).

[f] N. Nakagawa, *Nippon Kagaku Zasshi*, **82**, 141 (1961).

partial charge-transfer from the highest occupied π-orbital to the lowest unoccupied σ*-orbital of the CH bond is assumed to contribute to the stabilization of a CH/π interaction. If all other circumstances, especially steric conditions, were similar, the interaction would be facilitated by elevation of the HOMO (π) level and lowering of the LUMO (CH σ*) level.[28] From this point of view, electronic substituent effects were examined and compared with similar effects on typical hydrogen bonding[29] (Table 3.8).

In cases of intramolecular CH/π interaction, a favorable substituent effect should lead to an increase in the CH/π interacted conformer. In Chapter 2, the presence and abundance of these conformers were observable by various spectroscopic methods. The solution conformations are generally determined by IR, ^{1}H, and ^{13}C NMR spectra.[30]

The upfield shift δ_{CH} can be attributed to an increase in the CH/π interacted conformation: the CH group interacting with the aromatic nucleus suffers from a diamagnetic anisotropy effect by the ring current. Thus, the substituent-dependent upfield shift of the chemical shifts (δ_H) indicates the presence of a CH/π interacted conformer. In other cases, NOE enhancements of aromatic protons by the irradiation of the donor CH allowed estimation of the abundance of the CH/π interacted conformers. This effect is caused by the charge transfer (delocalization) character of the interaction,[31] and can be rationalized by the fact that the HOMO of a more electron-rich π-system donates electron more efficiently to the LUMO of a CH group.[32] However, the increase in electron density also favors the Coulombic interaction by increasing the negative charge of the π-system as well.

In the case of intermolecular interaction between chloroform and aromatic hydrocarbons, enthalpies of CH/π complex formation (Table 3.9) increases in the order of $C_6D_5Cl < C_6D_6 < p\text{-}C_6D_4(CD_3)_2 < C_6D_5CD_3 < o\text{-}C_6D_4(CD_3)_2 < C_6(CH_3)_6$.[33] Methyl

TABLE 3.9. Enthalpies and Entropies of CH/π Complex Formation in Ternary CHCl₃ (H-Donor)-Arene (H-Acceptor)-CCl₄ (Solvent) Systems

Arene	C_6D_5Cl	C_6D_6	C_6D_5Me	$C_6D_5Me_2$	C_6Me_6	$C_{10}D_8$
$-\Delta H$ (kcal mol^{-1})	1.52	1.89	2.09	2.21	2.98	1.73
ΔS (cal mol^{-1} K^{-1})	8.8	8.7	9.6	10.2	9.0	5.6

TABLE 3.10. The Electron Densities on CH Hydrogen Atoms and the Energies of the σ^*-Orbital of the CH Groups of Various CH Donors from PM3 Calculations and Their CH/π Interaction Enthalpies with C_6D_6

CH-Donor	ζ_H/e^a	E_{σ^*} (eV)	$-\Delta H$ (kcal mol^{-1})
CCl_3CHO	0.081	1.55	2.2
$CHBr_3$	0.138	2.71	2.0
$CHCl_3$	0.104	2.59	1.9
CH_2Cl_2	0.075	3.09	1.6
$CH(OCH_3)_3$	0.069	2.82	1.0

$^a e = 1.602 \times 10^{-19}$ C (charge of an electron).

groups donate electrons to the aromatic ring, while the chlorine atom is electron-withdrawing. The order agrees with the order of increasing electron density of the aromatic ring. The entropies of the chloroform–arene complex formation are within a relatively narrow range from -8.7 to -10.2 cal mol^{-1} K^{-1}, except for the case of naphthalene. Thus, the formation of the intermolecular CH/π complex is governed by the enthalpy of formation.

3.3.2. The Electronic Effects in CH Donors

The CH/π interaction is perturbed also by the electronic properties of the CH donor, in addition to the steric effects mentioned before. Table 3.10 lists the charge densities (ζ_H) and LUMO CH σ^*-orbital energies ($E\sigma^*$) of five CH donors together with the formation enthalpies of a CH/π complex with C_6D_6. The electronegative group on carbon attracts the CH bonding electron pair toward the carbon atom and increases the positive charge on H, as supported by the ζ_H values. The positive charge, in turn, strengthens the Coulombic interaction. The low $E\sigma^*$ of CCl_3CHO appears to increase the charge-transfer interaction.[34]

The conclusion of this section is summarized as follows: (1) electronegative substituents on the CH carbon atom intensify the CH/π interaction and (2) electron-donating substituents on the π-system favor CH/π interaction.

3.3.3. Steric Effects in the Intramolecular CH/π Interacted Systems

As in the case of intramolecular hydrogen bonds, the strength of intramolecular CH/π interaction and the relative abundance of the interacted species are affected by the topological distance (i.e., number of intervening bonds) between the CH and the π group ("ring size" effect) and also by the steric hindrance and conformational effects.

If we assume a priori that in alkylaromatics the CH group interacts with the *p*-orbital on the *ipso* carbon in counting the ring size (Fig. 3.8), the interaction forming a five-membered ring is the most favorable.[35] This trend agrees with many intramolecularly hydrogen-bonded systems and is thought to be entropically more favored than the longer chain homologs which form larger-membered rings.

Introduction of one or two methyl groups on an α-carbon atom always assists the CH/π interaction. However, the MM calculations showed that this comes, to a larger part, from the repulsive force betwen alkyl (CHRR' in Fig. 3.8) and methyl on α-carbon which

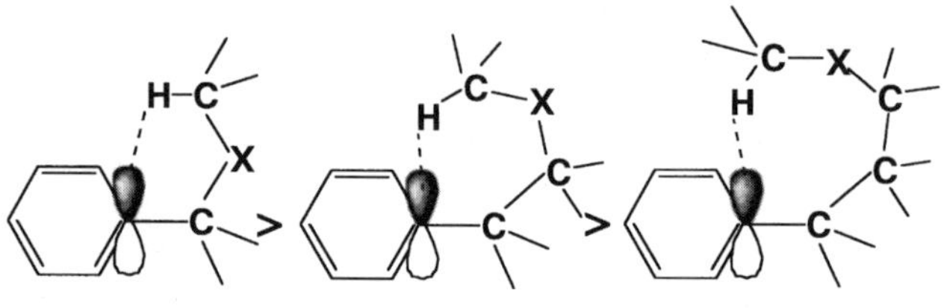

11 **12** **13**

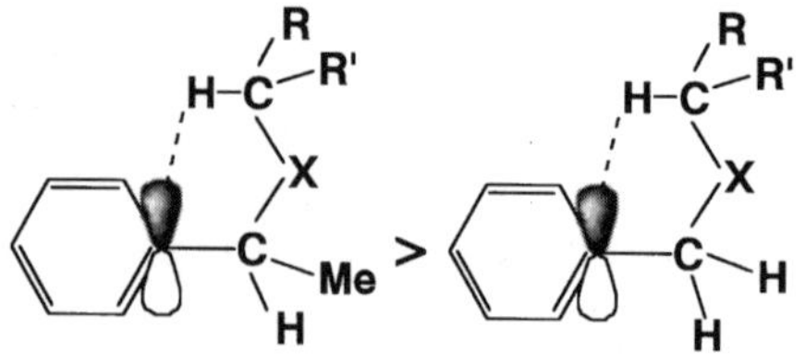

14 **15**

(more favorable) <---------> (less favorable)

Figure 3.8. Effects of the ring size and the introduction of α-alkyl group(s) on the CH/π interaction.

destabilizes the CH_3/CHRR' gauche conformer, and as a result, the conformer capable of CH/π interaction is favored. In short, CH/π interaction is sterically more favored in 1-phenylethyl derivatives than their benzyl analogs.[36]

3.4. NATURE OF THE CH/π INTERACTION

3.4.1. CH/π Interaction from the Viewpoint of the Hard–Soft–Acid–Base Concept

Using the hard–soft–acid–base principle,[37] Klopman[28] correlated the interaction between acids and bases to the LUMO energies of acids and the HOMO energies of bases. The LUMO energies of CH acids were obtained by MO calculations; those of methane and chloroform are $+0.8$ and -1.2 eV respectively. The values are within the range of a soft acid, as shown in Figure 3.9. The HOMO energies of aromatics are in the range from -6.61 eV (pentacene) to -9.25 eV (benzene),

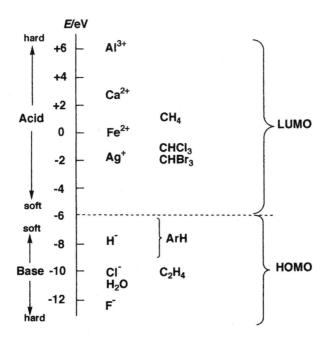

Figure 3.9. HOMO and LUMO energy levels of hard and soft acids and bases.

and those of ethylene, tetramethylethylene, and acetylene are -10.5, -8.27, together with the delocalized electron distribution in these molecules, and -11.4 eV, respectively.[38] These orbital energy data strongly support the fact that the CH/π interaction may be considered as a soft acid/soft base type, in which the Coulombic interaction is less important than the charge-transfer (delocalization) interaction. Klopman derived the following equation in order to evaluate the interaction energy between acid S and base R (whose MO coefficients are referred to as s and r, respectively,

$$\Delta E = - q_r q_s / r_{rs} + 2(C_{hr} C_{ls} \Delta \beta)^2 / (E_h - E_l) \tag{3.11}$$

wherein the suffixes h and l denote the HOMO of π-base and the LUMO of CH-acid, respectively.

In other words, the interaction should be frontier-controlled, which implies that the second term of Eq. 3.11 contributes to the stabilization of a CH/π interaction. The Coulombic term (the first term of Eq. 3.11), which controls the hard acid/hard base interaction, decreases sharply as the polarity of the solvent increases. Thus, the Na–Cl bond energy in the gaseous state is as high as 125 kcal mol^{-1}, but is dissociated easily in water, because of the high dielectric constant ($D = 84$ at $15°C$) of water which weakens the ionic bond drastically. In contrast, the orbital-controlled interaction term (the second term in Eq. 3.11) is not perturbed by an increase in the polarity of the solvent. This is reflected in the fact that the CH/π interaction is persistent in both nonpolar and highly polar solvents. This implies that the CH/π interaction can persist even in aqueous media and play an important role in many biological systems. Examples of CH/π interaction in biological molecules are given in Chapter 11.

3.4.2. Summary

In conclusion, the CH/π interaction can be described as a weak hydrogen bonding in which the dispersion contribution is relatively large. The features of CH/π interaction in comparison with other types of hydrogen bonding are summarized in Table 3.11.[39]

CH/π interaction is often confused with hydrophobic interaction. However, the former has many characteristics common to the hydrogen bond and can be considered as the weakest type of hydrogen bond.

TABLE 3.11. Comparison of CH/π Interactions with Other Similar Weak Hydrogen-Bond Interactions

Type of Weak Bond[a]	Interaction Energy (kcal/mol)	Type of Interaction			
		Delocalization	Electrostatic	Dispersion	Repulsive van der Waals
O–H(HA) ··· O(HB)	10–3	Variable	Strong	Unimportant	Similar
O–H(HA) ··· π(SB)	ca. 2	Important	Weak	Important	
C–H(SA) ··· O(HB)	< 3	Unimportant	Important	Unimportant	
C–H(SA) ··· π(SB)	**< 2.5**	**Important**	**Unimportant**[b]	**Important**	

[a] HB, SA, and SB refer to hard and soft acids and bases, respectively.
[b] In the case of rather strong CH acids which bear more than one electronegative group on their carbon atoms, electrostatic interaction is expected to be considerably important.

REFERENCES

1. P. Hobza and R. Zahradnik, *Weak Intermolecular Interactions in Chemistry and Biology*, Elsevier, Amsterdam, 1980.
2. P. Arrighini, *Intermolecular Forces and their Evaluation by Perturbation Theory*, Springer Verlag, Berlin, 1981.
3. A. Streitwieser, *Molecular Orbital Theory for Organic Chemists*, Wiley, New York, 1961, pp. 33–39; J. A. Pople and D. L. Beveridge, *Approximate Molecular Orbital Theory*, McGraw-Hill, New York, 1970, p. 11.
4. W. J. Hehre, *Acc. Chem. Res.*, **9**, 399 (1976); J. L. Whitten, *Acc. Chem. Res.*, **6**, 238 (1973).
5. W. G. Richards and D. L. Cooper, *Ab initio Molecular Orbital Calculations for Chemists*, 2nd ed., Clarendon, Oxford, 1983.
6. K. Morokuma, *Acc. Chem. Res.*, **10**, 294 (1977); *J. Chem. Phys.*, **58**, 5823 (1973); K. Kitaura and K. Morokuma, *Int. J. Quantum Chem.*, **10**, 325 (1976).
7. K. Morokuma and G. Wipff, *Chem. Phys. Lett.*, **74**, 400 (1980).
8. P. A. Kollman and L. C. Allen, *J. Chem. Phys.*, **52**, 5085 (1970).
9. K. Nikki and N. Nakagawa, *Chem. Lett.*, 699 (1974); K. Nikki, N. Nakahata and N. Nakagawa, *Tetrahedron Lett.*, 3811 (1975); K. Nikki, N. Nakagawa and Y. Takeuchi, *Bull. Chem. Soc. Jpn.*, **48**, 2902 (1975).
10. F. London, *Z. Phys.*, **63**, 245 (1930); *Z. Phys. Chem.*, **11**, 222 (1930).
11. Dispersion force second-order perturbation equation.
12. G. Klopman, *J. Am. Chem. Soc.*, **90**, 223 (1968); L. Salem, *J. Am. Chem. Soc.*, **90**, 543, 553 (1968).
13. C. A. Coulson, *Hydrogen Bonding*, D. Hadzi and H. W. Thompson, Eds., Pergamon Press, Oxford, 1959, p. 339.
14. S. Araki, T. Seki, K. Sakakibara, M. Hirota, and M. Nishio, *Tetrahedron: Assym.*, **4**, 555 (1993).
15. Review: P. Laszlo, *Progr. Nucl. Magn. Reson.*, **3**, 348 (1967); R. F. Zurcher, *Helv. Chim. Acta*, **44**, 1380 (1961); *ibid.*, **46**, 2054 (1963); N. S. Bhacca and D. H. Williams, *Tetrahedron Lett.*, 3127 (1964); D. H. Williams and N. S. Bhacca, *Tetrahedron*, **21**, 2021 (1965); P. Diehl, *J. Chem. Phys.*, **61**, 179 (1964); T. Ledaal, *Tetrahedron Lett.*, 1683 (1968); W. G. Schneider, *J. Phys. Chem.*, **66**, 2653 (1962); T. L. Brown and K. Slark, *J. Phys. Chem.*, **69**, 2679 (1965); R. Ehama, A. Yokoo, M. Tsushima, T. Yuzuri, H. Suezawa, and M. Hirota, *Bull. Chem. Soc. Jpn.*, **66**, 814 (1993).
16. S. L. Price and A. J. Stone, *J. Chem. Phys.*, **86**, 2859 (1987).
17. A. D. Buckingham and P. W. Fowler, *J. Chem. Phys.*, **79**, 6426 (1983).

18. J. Pawliszyn, M. M. Szezesniak, and S. Scheiner, *J. Phys. Chem.*, **88**, 1726 (1984).

19. T. Aoyama, O. Matsuoka, and N. Nakagawa, *Chem. Phys. Lett.*, **67**, 508 (1979).

20. Y. Kodama, K. Nishihata, M. Nishio, and N. Nakagawa, *Tetrahedron Lett.*, 2105 (1977); M. Nishio, *Kagaku no Ryoiki*, **33**, 422 (1979).

21. T. Takagi, A. Tanaka, S. Matsuo, H. Maezaki, M. Tani, H. Fujiwara, and Y. Sasaki, *J. Chem. Soc.*, *Perkin 2*, 1015 (1987).

22. M-F. Fan, Z. Lin, J. E. McGrady, and D. M. P. Mingos, *J. Chem. Soc.*, *Perkin 2*, 563 (1996).

23. K. Morokuma and G. Wipff, *Chem. Phys. Lett.*, **74**, 400 (1980).

24. M. Oki, T. Onoda, and H. Iwamura, *Tetrahedron*, **24**, 1905 (1968).

25. K. Abe, M. Hirota, and K. Morokuma, *Bull. Chem. Soc. Jpn.*, **58**, 2713 (1985).

26. S. Araki, K. Sakakibara, M. Hirota, M. Nishio, S. Tsuzuki, and K. Tanabe, *Tetrahedron Lett.*, 6587 (1991).

27. N. L. Allinger and H. L. Flanagan, *J. Comput. Chem.*, **4**, 399 (1983).

28. G. Klopman, *Chemical Reactivity and Reaction Paths*, G. Klopman, Ed., Wiley-Interscience, New York, 1974, p. 55.

29. G. C. Pimentel and A. L. McClellan, *The Hydrogen Bond*, Freeman, San Francisco, CA, 1960; D. Hadzi and H. W. Thompson, *Hydrogen Bonding*, Pergamon Press, London, 1959; L. N. Ferguson, *Modern Structural Theory of Organic Chemistry*, Prentice-Hall, Englewood Cliffs, NJ, 1963.

30. I. Solomon, *Phys. Rev.*, **99**, 559 (1955); N. Bloembergen, *J. Chem. Phys.*, **27**, 572 (1957); E. D. Becker, *High Resolution NMR Theory and Chemical Applications*, 2nd ed., Academic Press, New York, 1980, Chapter 9; C. C. Hinckley, *J. Am. Chem. Soc.*, **91**, 5160 (1969); *J. Org. Chem.*, **35**, 2834 (1970); A. F. Cockeril, G. L. O. Davies, R. C. Harden, and D. M. Rackham, *Chem. Rev.*, **73**, 553 (1973).

31. Several examples are found in M. Oki and H. Iwamura, *Bull. Chem. Soc. Jpn.*, **32**, 113 (1959); M. Oki and M. Hirota, *Spectrochim. Acta*, **17**, 583 (1961).

32. V. Gutmann, *The Donor–Acceptor Approach to Molecular Interactions*, Plenum, New York, 1978.

33. R. Ehama, A. Yokoo, M. Tsushima, T. Yuzuri, H. Zuezawa, and M. Hirota, *Bull. Chem. Soc. Jpn.*, **66**, 814 (1993).

34. G. Klopman, *J. Am. Chem. Soc.*, **90**, 223 (1968).

35. H. Suezawa, R. Ehama, T. Hashimoto, and M. Hirota, unpublished data.

36. M. Hirota, K. Abe, T. Sekiya, H. Tashiro, M. Nishio, and E. Osawa, *Chem. Lett.*, 685 (1981); M. Hirota, T. Sekiya, K. Abe, H. Tashiro, M. Karatsu, M. Nishio, and E. Osawa, *Tetrahedron*, **39**, 3091 (1983).

37. R. G. Pearson, *J. Am. Chem. Soc.*, **85**, 3533 (1963); *J. Chem. Educ.*, **45**, 58 (1968); T. L. Ho, *Hard Soft Acids and Bases Principle in Organic Chemistry*, Academic Press, New York, 1977.

38. Kagaku Benran, Ed., *Handbook of Chemistry*, Chemical Society of Japan, Maruzen, Tokyo, 1994.

39. M. Nishio and M. Hirota, *Tetrahedron*, **45**, 7201 (1989).

CHAPTER 4

COMPARISON WITH OTHER TYPES OF WEAK MOLECULAR INTERACTIONS

4.1. CH/Y HYDROGEN BONDS

If CH/π interaction is hydrogen-bond-like, the CH group should be able to act as a hydrogen donor to the other acceptors of hydrogen bonding. The CH/Y (Y: electronegative atom, mostly O or N) interaction[1,2] will be briefly reviewed in comparison with the CH/π interaction.

4.1.1. Intermolecular CH/Y Interaction in Solutions

Hydrogen bonding of chloroform and other haloforms with oxygen-centered hydrogen acceptors has long been known. Before the concept of hydrogen bonding came into chemistry, the association of chloroform with acetone was suspected.[3] The ability of CH groups as hydrogen donors was critically surveyed by Allerhand and Schleyer.[4] They measured the infrared CH absorptions of a large number of CH donors in the presence and absence of hydrogen acceptors. In carbon tetrachloride solutions containing strong hydrogen acceptors, namely dimethyl-d_6 sulfoxide and pyridine-d_5 (Fig. 4.1), CH stretching absorption bands of the CH/O and CH/N complexed species appeared at frequencies lower than that of the free species.

The IR absorptions of CH-donor/hydrogen bond acceptor/CCl_4 systems are shown in Table 4.1.[5] Haloforms and other tris(negatively substituted)methanes form CH/Y (Y = O, N) interacted complexes. Judging from the hydrogen bond shift ($\Delta v = v_{free} - v_{interacted}$), the strength of CH/Y hydrogen bond decreases in the order: $CHCl_3 > CHI_3 > CHBr_3$. Some bis(negatively substituted)methanes, such as chloroacetonitrile, can be hydrogen donors to oxygen-centered acceptors. The effectiveness of electronegative groups in activating the CH donor decreases in the following order (by IR Δv

X,Y,Z; Electronegative groups

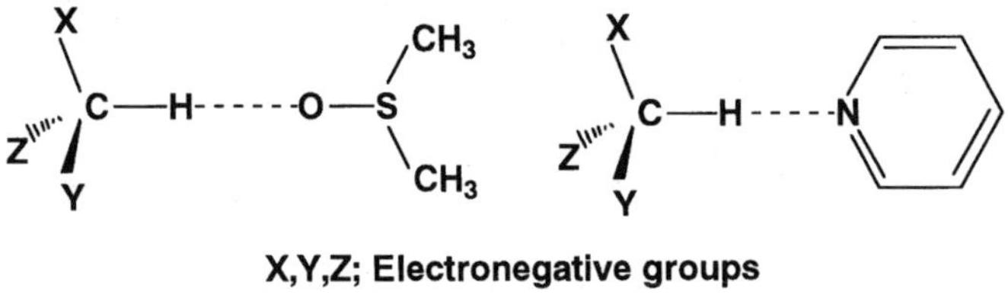

Figure 4.1. Intermolecular CH/O and CH/N hydrogen bonding.

TABLE 4.1. CH/O and CH/N Hydrogen Bonding of Various CH-Donors with Dimethyl-d_6 Sulfoxide and Pyridine-d_5 (in CCl_4 Solutions)

CH-Donor	v_{free} (cm^{-1})	Δv (cm^{-1}) 0.95 M DMSO	Δv (cm^{-1}) 2 M Pyridine
Br_2CHCN	2998	80	113
Cl_2CHCN	2987	66	92
$CHBr_2CBr_3$	2993	58	87
$CHBr_3$	3031	50	82
$CHClBr_2$	3027	43	—
$CHCl_2CCl_3$	2986	41	64
CHI_3	3011	40	55
$CHCl_3$	3020	29	46
$CHCl=CCl_2$	3084	41	45
$1,2,4,5-C_6H_2Cl_4$	3092	40	42
ICH_2CN	2968	23	—
$ClCH_2CN$	2966	15	30
$BrCH_2C{\equiv}CH$	3311	102	110
$CH_3(CH_2)_3C{\equiv}CH$	3315	82	—

TABLE 4.2. The CH Hydrogen Bonding Between 1-Octyne and Hydrogen Acceptors

H-Acceptor	K (1 mol^{-1} at 25°C)	$-\Delta H$ (kcal mol^{-1})
Acetone	0.14	1.5
Acetonitrile	0.16	1.8
Pyridine	0.21	2.0

criterion): $CN > CBr_3 > CCl_3 > Br > I > Cl > COOR > CHBr_2 > CHCl_2$. The effect of fluorine was inconclusive because of the high volatility of fluorine-containing compounds.

A proton attached to a sp^2-hybridized carbon can also form complexes when the molecule has an electronegative substituent such as a halo or cyano group.

Acetylenic hydrogen atoms (on a sp-carbon) are more acidic than sp^2- and sp^3-hybridized CHs and are stronger donors in the CH/Y hydrogen bond. 1-Octyne interacts with some hydrogen acceptors even if the CH group has no electronegative substituents. The interaction is strong, as shown by the large equilibrium constants (K) and enthalpies (ΔH) of CH/O and CH/N complex formation given in Table 4.2.[6]

Calculations on the CH/O hydrogen bonds showed that the contribution of charge-transfer from a nonbonding orbital of an O atom to σ^*(CH), $2s$(H), or $2p$(H)-orbital is insignificant and that the major stabilization arises from the electrostatic interaction energy.[7]

4.1.2. Intermolecular CH/Y Interactions in Solids

The CH/Y distances known from X-ray and neutron-diffraction data can be useful probes to determine CH/Y hydrogen-bonded pairs of atoms in the crystalline state. Taylor and Kennard[8] surveyed neutron diffraction data on 113 compounds accumulated in the Cambridge Structural Database,[9] and showed that many hydrogen atoms covalently bonded to carbon tend to form short intermolecular contacts to oxygen atoms rather than to carbon or hydrogen atoms. They assumed that the distance ($r_{H \cdots Y}$) between the hydrogen atom (H) and the acceptor atom (Y = C, H, O, N, Br, Cl, P, or S) is shorter than the sum of their van der Walls radii[10] [R(H) and R(Y), respectively]; that

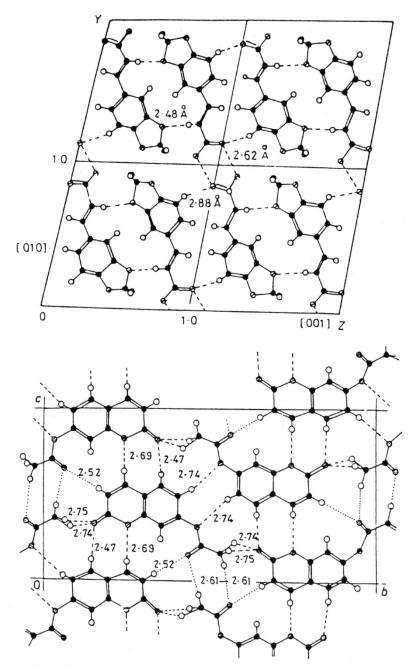

Figure 4.2. Two examples of X-ray crystal structure containing CH/O hydrogen bonds. Upper; 3,4-methylenedioxycinnamic acid. Lower; 7-acetoxycoumarin.

is, d value is positive when a CH/Y hydrogen bond exists (Eq. 4.1).

$$d = R(H) + R(Y) - r_{H \cdots Y} \qquad (4.1)$$

Contacts with C–H $\cdots$ Y angles of less than 90° were ignored, taking the geometrical requirement for hydrogen-bond formation into account. Their survey showed that the (C–)H $\cdots$ O pairs are 2.5 times more likely to show positive d values than was statistically expected. Moreover, out of 19 intermolecular (C–)H $\cdots$ Y contacts with $d > 0.3$ Å, 15 (79%) are (C–)H $\cdots$ O pairs.

Short CH $\cdots$ O contacts have been reported by many other crystallographic investigations as well. Some examples are given in Fig. 4.2,[11] in which we see that not only hydrogen atom attached to olefinic and aromatic sp^2-carbons, but also methyl hydrogen atoms, can be located at close distances to the carbonyl and the ether oxygen atoms.

4.1.3. Intramolecular CH/Y Interaction

The ability of the formyl group as a CH donor in CH/Y hydrogen bonding has been controversial. Several papers for and against the presence of formyl CH/O hydrogen bonding have been published.

Pinchas found an intramolecular CH/O hydrogen bond in benzaldehydes that bear hydrogen-accepting substituents at the ortho position(s).[12,13] However, in sharp contrast to the usual OH and NH hydrogen bonds, infrared CH absorption bands were observed to shift to higher frequencies. In the case of o-nitrobenzaldehyde **1**, the O atom approaches proton from the direction of the acute C–H $\cdots$ O angle (Fig. 4.3). However, Pinchas explained the high-frequency shift by the steric compression effect. His interpretation was criticized by Forbes on the basis of spectral data.[14] Rao showed that a conformation with the NO_2 group 90° out of the aromatic plane is the most stable by EHT + CNDO2 calculation.[6] Thus, the usual CH/O hydrogen bond using the lone-pair electrons on the oxygen is impossible. However, a CH/π bond is more likely in this conformation.

Molecular orbital studies on the CH/O hydrogen bonding of formylthiophenecarboxylate esters gave important information concerning the nature of the CH/O interaction.[15] Of all the positional isomers, methyl 2-formyl-3-thiophenecarboxylate **2** in a CH/O-proximate conformation has the highest s-character and highest positive charge in its formyl CH bond and the highest negative charge on the carbonyl

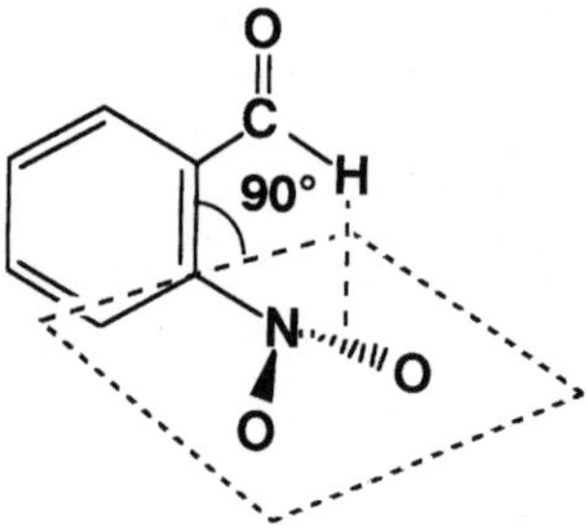

1

Figure 4.3. The most stable conformation of *o*-nitrobenzaldehyde suggested by Rao et al.[6] on the basis of MO calculations.

TABLE 4.3. Some MO Properties of Selected Isomers of Methyl Formylthiophene-2-carboxylates (by MNDO)

Compound	*s*-Character (CHO) from $^1J_{CH}$ (calcd.)	q_H (CHO)	q_O (ester C=O)	$P_{O \cdots H}$ (nonbonded)
2	38.3 (39.0)	0.071	− 0.387	0.049
3	36.2 (38.0)	0.038	—	0.000
4	37.5 (38.5)	0.047	− 0.379	0.047

oxygen (only three are given in Table 4.3). The high *s*-character increases the polarity of the CH bond and favors Coulombic interaction between 2-formyl hydrogen and 3-carbonyl oxygen. In addition, it has the highest bond population between the nonbonded H(formyl) and O(carbonyl) atoms ($p_{O \cdots H}$ in Table 4.3) among all the possible conformers of all the positional isomers. These results suggest the presence of weak delocalization interaction between these atoms.

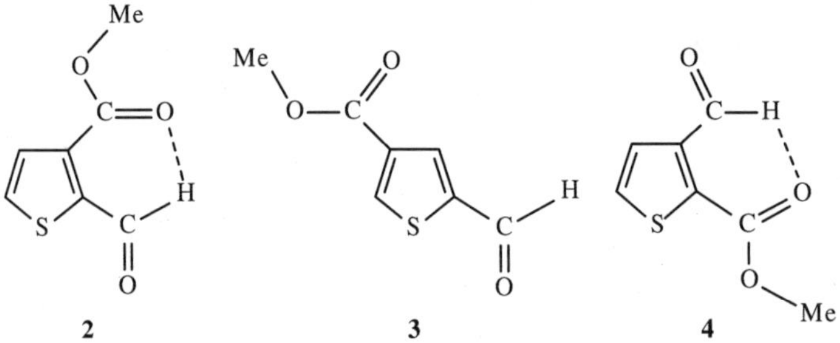

2   **3**   **4**

Another example of a strong CH/O hydrogen bond was reported on 1-trifluoroacetoxy- and 1,3-bis(trifluoroacetoxy)-2-hydrylperfluoro-adamantanes (**6** and **7**).[16] In this case, the CH group is activated by a highly electronegative perfluoroalkyl group.

Infrared CH frequencies (v_{CH}) and the proton chemical shifts (δ_{CH}) of CH groups capable of forming CH/O hydrogen bonds are collected in Table 4.4. The nonpolar CH bond can manifest its proton-donating ability only in interactions with strong proton acceptors in sterically favorable circumstances. In all cases, downfield shifts of CH-donor protons were observed.

A case of CH/O hydrogen bond strong enough to enforce a molecule of tricyclic orthoamide **8** to take an eclipsed conformation was reported by Seiler and Dunitz[17] based on an X-ray crystallographic study. A trihydrate crystal of **8** was shown to take a nearly eclipsed conformation (N–C–C–H torsional angle; 8.2°) about the central C–CH$_3$ bond (Fig. 4.4). In the crystal, the three oxygen atoms of

TABLE 4.4. Effects of CH Hydrogen Bonding on the IR v_{CH} Absorptions and ^{1}H-NMR Chemical Shifts in Several Intramolecular CH/O Hydrogen-Bonded Molecules

Compound	v_{CH} (cm^{-1})	δ_H (ppm) ($^1J_{CH}$/Hz)
o-Nitrobenzaldehyde	2650, 2760, 2893	
p-Nitrobenzaldehyde	2725, 2865	
2-Formylthiophene-3-COOMe **2**	2896	10.65 (192)
5-Formylthiophene-3-COOMe **3**	2825	9.87 (181)
3-Formylthiophene-2-COOMe **4**	2904, 2881	10.62 (189)
4-Formylthiophene-2-COOMe	2789, 2804, 2845	9.90 (185)
2-H-F-adamantane **5**	2986	5.45
1-CF$_3$COO-2-H-F-adamantane **6**	3020	6.55
1,3-Di(CF$_3$CO$_2$)-2-H-F-adamantane **7**	3051	7.70

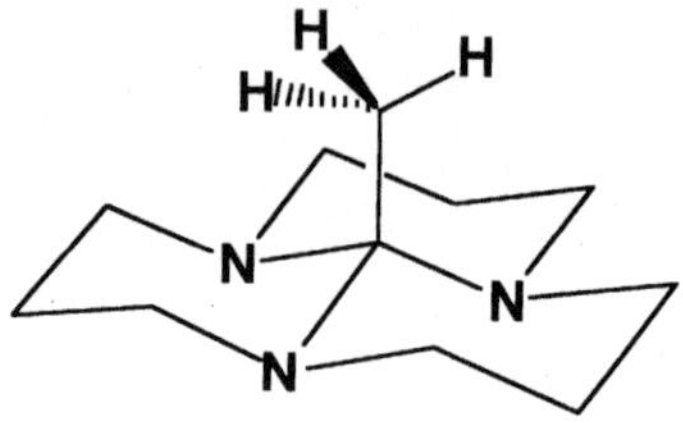

8

Figure 4.4. The H/N eclipsed conformation of the tricyclic orthoamide.

hydrate water molecules are located opposite the three methyl protons (N–C–C $\cdots$ O torsional angle; 0.9°). The driving force responsible for this eclipsed conformation is thought to be intermolecular CH/O hydrogen bonding.

4.2. XH/π INTERACTIONS

Historically the concept of CH/π interaction was proposed after the OH/π and NH/π interactions had been found in organic chemistry. In this respect, the CH/π interaction is an extension of the series to its weakest limit.

4.2.1. Intermolecular XH/π Interactions

It has long been known that the π-electrons of alkenes, alkynes, and aromatic compounds act as hydrogen acceptors.[18,19] Alcohols and phenols are known as good hydrogen donors, as well as primary and secondary amines, to π-bases. Interaction between π-bases and hydrogen donors was examined mainly by IR spectroscopy.[20,21] Quite similar to the more typical hydrogen bonds involving the lone-pair electrons of electronegative atoms as proton acceptor, the OH and NH bonds were shown to be considerably weakened. As a consequence, the OH and NH stretching absorptions of hydrogen donors show low-frequency shifts.

   Infrared v_{XH} spectra of XH/π hydrogen-bonded systems were extensively studied by Josien et al. (Table 4.5).[22] The frequency shift of XH absorption in dilute carbon tetrachloride solution is known as the hydrogen bond shift Δv. The XH frequencies (v_{XH}) and the Δv values of various XH/π complexes are collected in Table 4.5. The low-frequency shifts due to hydrogen bond formation originate from the lowering

TABLE 4.5. X–H Stretching Frequencies (cm^{-1}) and the Hydrogen Bond Shifts ($\Delta\nu$/cm^{-1}) due to XH/π Hydrogen Bond Formation[22]

						H-Donor			
	HCl	HBr	HI	CH$_3$OH	C$_6$H$_5$OH	Pyrrole	MeCONHMe	C$_6$H$_5$SH	CDCl$_3$
Gas	2866	2558	2230	3687	3654	3570	—	2592	2263.5
CCl$_4$	2831	2518	2204	3645	3610	3495	3431	2591	2252
C$_6$H$_5$Cl	2779	2481	2182	3624	3578	3475	—	—	2249
	(52)	(37)	(22)	(21)	(32)	(20)			(3)
C$_6$H$_6$	2750	2476	2164	3607	3557	3460	3406	2574	2247.5
	(81)	(42)	(40)	(38)	(53)	(35)	(25)	(17)	(4.5)
C$_6$H$_5$CH$_3$	2735	2452	2157	3607	3547	3451	3400	—	2247
	(96)	(66)	(47)	(38)	(63)	(44)	(31)		(5)
m-C$_6$H$_4$(CH$_3$)$_2$	2723	2430	2147	3593	3538	3447	3395	—	2247
	(108)	(88)	(57)	(52)	(72)	(48)	(36)		(5)
Mesitylene	2712	2416	2132	3593	3530	3440	3389	2564	2246.5
	(119)	(102)	(72)	(52)	(80)	(55)	(42)	(27)	(5.5)
Durene	2707	2409	2125	3582	3523	3435	3384	—	—
	(124)	(109)	(79)	(63)	(87)	(60)	(47)		
C$_6$(CH$_3$)$_6$	2676	2380	2097	3570	3505	3418	—	—	2244
	(155)	(138)	(107)	(75)	(105)	(77)			(8)

TABLE 4.6. Substituent Effect on the Interaction Enthalpies and the Equilibrium Constants of Intermolecular XH/π Interacted Systems[a]

π-Bases	CH Acids[b]								OH Acid[c]	NH Acid[d]
	CHCl₃[b]		CHBr₃	CH(OMe)₃	CH₂Cl₂	CCl₃CHO			C_6H_5OH	$CH_3CONHCH_3$
	K_{298}	$-\Delta H$	K_{298} ($-\Delta H$)	K_{298} ($-\Delta H$)	K_{298} ($-\Delta H$)	K_{298} ($-\Delta H$)			K_{302}	K_{298}
C_6H_5Cl	0.18	1.52								
C_6H_6	0.30	1.89	0.505 (1.99)	6.30 (0.99)	0.362 (1.59)	0.605 (2.18)			0.33	0.32
$C_6H_5CH_3$	0.27	2.09							0.39	0.32
$C_6H_5C_2H_5$									0.35	
$o\text{-}C_6H_4(CH_3)_2$	0.25	2.21							0.35	0.36
$p\text{-}C_6H_4(CH_3)_2$	0.28	2.04							0.39	0.36
Mesitylene	0.38								0.57	0.48
Durene	0.43								0.60	0.44
$C_6(CH_3)_6$	1.62	2.98							0.63	
Naphthalene	1.07	1.73							0.58	0.56

[a] In the experiments on CH acids, deuterium-labeled aromatic compounds were used as π-bases.
[b] Reference 23.
[c] Z. Yoshida and E. Osawa, *Nippon Kagaku Zasshi*, **87**, 509 (1966).
[d] T. Shimanouchi, M. Tsuboi, and I. Suzuki, *J. Chem. Phys.*, **31**, 1437 (1959); *ibid.*, **32**, 1263 (1960).

of the force constant of the XH bond. As expected, for XH donors in Table 4.5, the lower frequency shifts increase as the basicity, or electron-donating ability increases, but they are considerably smaller than in OH/O and other hydrogen bonds. For comparison, Δv of methanol with diethyl ether and pyridine are 151 and 304 cm^{-1}, respectively. Phenol is a stronger H-donor, and its Δv values for the same solvents are 275 and 468 cm^{-1}, respectively. Chloroform-d in Table 4.5 is an example of CH(D)-donor. The C–D frequency of $CDCl_3$ is rather insensitive to the π-basicity of aromatic compounds, but careful examination of a series of CH/π complexes revealed that the CH shifts also follow the regular rule of π-basicity.

The formation constants and the enthalpies of formation of XH/π complexes are given in Table 4.6. For comparison, those for the CH/π interacted system ($CDCl_3$ as the H-donor)[23] are given together.

The formation constants for OH/π and CH/π interactions with the same π-base are similar in spite of the large difference between their hydrogen bond shift Δv. In contrast, both the formation constant and the enthalpy of formation of a strong hydrogen bond are considerably larger than those for CH/X interactions (cf. K for the phenol–pyridine complex is as high as 55).[24] This is quite similar to CH/X hydrogen bonds, as discussed in Section 4.1, where the hydrogen bond shift is also small as in the case of CH/π interaction.

The presence of a hydrogen-bond interaction between ethylene and water was detected in an argon matrix.[25] It was shown that the water molecules exist as dimers and that the hydrogen-accepting water molecules behave in turn as the hydrogen donor to ethylene (Fig. 4.5).

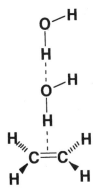

Figure 4.5. Hydrogen-bonded water–ethylene complex detected in an argon matrix.

4.2.2. Intramolecular XH/π Interactions

As in the case of hydrogen bonding, strong proton donor groups such as OH, NH, and SH are expected to participate in intramolecular interaction with a π-group within the same molecule. Such an intramolecular interaction is strictly dependent on the steric requirements to maintain the geometry of the interacting system. The characteristic feature and the limitations of intramolecular OH/π interaction has been systematically studied by Ōki and Iwamura.[26] The extent of the steric limitations for the persistence of intramolecular OH/π interaction are given in Table 4.7. An example of an intramolecular NH/π interacted system[27] is given for comparison.

TABLE 4.7. Extent and Limitations of Intramolecular XH/π Interaction

XH/π Molecules	n^a	Ring Size
X = O		
$CH_2=CH-(CH_2)_nOH$ **9**	1, 2, (3)	5, 6, (7)
cyclo-$C_3H_5-(CH_2)_nOH$ **10**	1, 2	5, 6
$HC\equiv C-(CH_2)_nOH$	1, 2, 3	5, 6, 7
$C_6H_5-(CH_2)_nOH$ **11**	1, 2, 3	5, 6, 7^a
$o\text{-}[CH_2=CH-(CH_2)_n]C_6H_4OH$	0, 2	5, 6
$o\text{-}[HC\equiv C-(CH_2)_n]C_6H_4OH$	0, 1, - - -^{c}	5, 6, - - -
$o\text{-}[C_6H_5(CH_2)_n]C_6H_4OH$	0, 1, 2	5, 6, 7^d
X = N		
$C_6H_5-(CH_2)_nNHC_6H_5{}^b$	1, 2, 3, 4, (5)	5, 6, 7, 8, (9)

^{a} The chain-length n requisite for the intramolecular interaction.
^{b} In the cases of long-chain N-(ω-phenylalkyl)anilines, the formation of intramolecular charge-transfer complexes are suspected.
^{c} The compounds with more than two n values were not examined.
^{d} Ring size assuming that the interaction takes place at the *ipso* position.

Quite similar to the usual OH $\cdots$ O and OH $\cdots$ N hydrogen bonding, the interactions forming five- and six-membered rings are most favorable. Moreover, the seven-membered ring is also allowed to persist in the interaction in some cases. A similar trend is also seen with the CH/π interacted systems.

4.2.3. OH/π Interactions in Solids (Evidence from Crystallography)

Intermolecular OH/π proximate arrangements in crystals are usual and were reported in several papers. Interaction of the OH group of 2,2,2-trifluoro-1-(9-anthryl)ethanol with the π-face of another molecule in a crystal was shown by X-ray crystallography.[28] X-ray studies on crystalline Na_4[calix[4]arenesulfonate] $\cdot 13.5H_2O$[29] showed that the hydrophobic cavity was formed by aromatic groups and it contained a water molecule when the hydrophobic guest molecule is absent. Two hydrogen atoms of the water molecule form OH/π hydrogen bonds with two of the aromatic rings of the calixarene skeleton (Fig. 4.6). The results gave us an insight into the mechanism of the interactions of aromatic nuclei in biomolecules (phenylalanine, tyrosine, and others).

In the X-ray crystal structure of 4-nitro-2,6-diphenylphenol **12** reported by Ueji et al., the hydroxyl group is bifurcated and involved in both intra- and intermolecular hydrogen bonds.[30] The finding was supported by IR spectroscopy. The OH proton is located just above one C–C bond (and not just above the center) of both the intra- and intermolecular acceptor benzene rings [C(1')–C(2') and C(3″)–C(4″) bonds, respectively] (see Fig. 4.7). These observations provide evidence for the participation of the delocalization (charge-transfer) force in the OH/π interaction.

$$O_2N-\overset{\overset{\displaystyle C_6H_5}{|}}{\underset{\underset{\displaystyle C_6H_5}{|}}{\bigcirc}}-OH$$

12

The intramolecular OH/aryl interacted structure of 2,2-bis(2-hydroxy-5-methyl-3-t-butylphenyl)propane **13** is the first evidence of

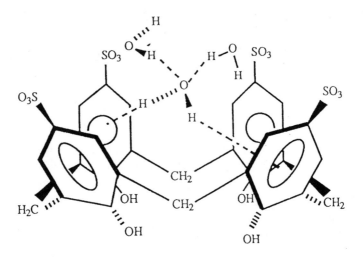

Figure 4.6. Water molecules in the hydrophobic cavity of a calixarene molecule.

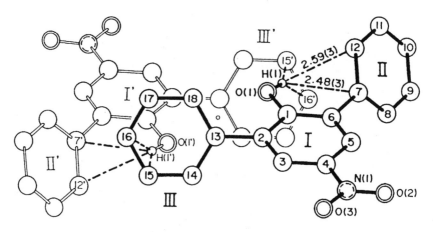

Figure 4.7. Intra- and intermolecular OH/π interactions in the crystal of 4-nitro-2,6-diphenylphenol.

OH/π interaction from X-ray crystallography.[31] The distance between the hydroxyl proton and the π-plane of the aryl group is 2.09 Å. Intramolecular OH/π interaction in crystals of similar bis(hydroxyphenyl)alkanes related to novolaks has been confirmed by IR spectroscopy.[32]

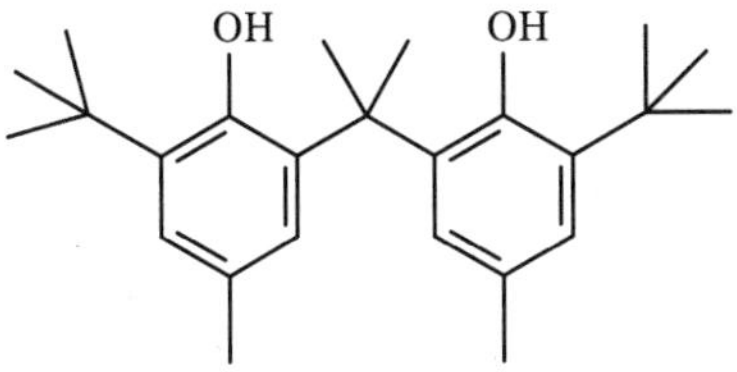

13

Organosilanols generally show a strong tendency to associate inter-molecularly through OH···O hydrogen bonds in the solid state. However, an interesting example of a silanol OH group interacting with a phenyl group in the same molecule was recently reported in the crystal of a sterically crowded [tris(dimethylphenylsilyl)methyl]-methylsilanol **14**.[33]

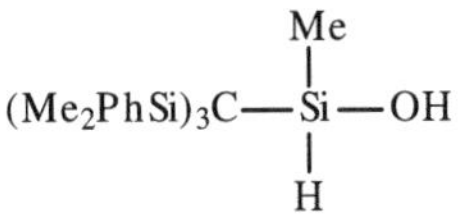

$$(Me_2PhSi)_3C \!-\! \underset{\underset{H}{|}}{\overset{\overset{Me}{|}}{Si}} \!-\! OH$$

14

In anti-10-endo-hydroxy-10-exo-butyltricyclo[4.2.1.1]deca-3,7-dien-9-one **15**,[34] the 10-endo-hydroxyl proton is found as close as 2.11 and 2.16 Å to the two olefinic carbon atoms [C(7) and C(8), respectively]. The presence of intramolecular OH/π interaction was further proved by the appearance of ν_{OH} at 3600 cm^{-1} in CCl$_4$ solution.

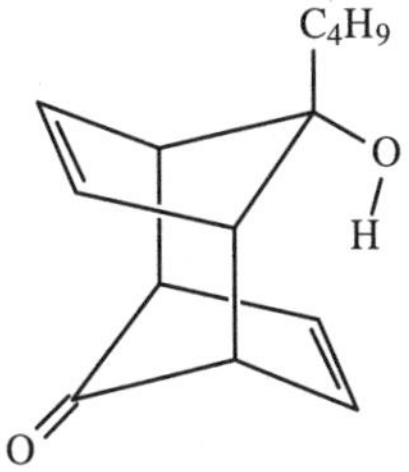

15

4.3. WEAK COULOMBIC INTERACTIONS

Coulombic interactions can be expressed as the sum of the attraction terms between charges and/or permanent multipoles. The interaction

TABLE 4.8. Various types of Coulombic Interactions

Coulombic Interaction Subsystem		Type of Interaction	E_C Term
R	S		
(+)	(−)	Charge-charge (ion) (ion)	$\dfrac{1}{4\pi\varepsilon_0}\dfrac{q_R q_S}{r}$
(+)	(− +)	Charge-dipole	$\dfrac{1}{4\pi\varepsilon_0}\dfrac{kq_R\mu_S}{r^2}$
(− +)	(− +)	Dipole-dipole	$\dfrac{1}{4\pi\varepsilon_0}\dfrac{k\mu_R\mu_S}{r^3}$
(+)	(− + + −)	Charge-quadrupole	$\dfrac{1}{4\pi\varepsilon_0}\dfrac{kq_R Q_S}{r^3}$
(− +)	(− + + −)	Dipole-quadrupole	$\dfrac{1}{4\pi\varepsilon_0}\dfrac{k\mu_R Q_S}{r^4}$
(− + + −)	(− + + −)	Quadrupole-quadrupole	$\dfrac{1}{4\pi\varepsilon_0}\dfrac{kQ_R Q_S}{r^5}$

between two net electron charges is as strong as a covalent bond in a vacuum or in a nonpolar medium. Interactions weaker than dipole–dipole interactions are usually classified as weak. Various Coulombic interactions are given in Table 4.8,[35] where q, μ, and Q refer to charge, dipole moment, and quadrupole moment, respectively. The magnitude of interaction tends to decrease on going down the table. The range of the interaction becomes shorter and shorter (or near-reaching) in the same order.

All Coulombic interaction terms tend to decrease in a manner that is inversely proportional to the dielectric constant of the medium. Thus, this sort of weak interaction will be obscured in polar solvents and, hence, are observable only in gaseous state or in solutions of nonpolar solvents. In the case of the CH/π interaction, π-base is generally not dipolar. Even if the C–H group can be a strong dipole when the carbon atom is substituted with more than one electronegative groups, the interaction cannot be so strong as to be observable in polar solvents. Actually CH/π interactions are shown to persist even in polar solvent. Thus, the CH/π interaction is clearly different from Coulombic interaction. [The aromatic π-system has a large quadrupole which can enhance the CH/π interaction nevertheless.]

4.4. HYDROPHOBIC INTERACTION

Hydrophobic interaction operates between nonpolar organic molecules and water. Since Kauzmann[36] reported that the hydrocarbon parts of amino acid residues in a polypeptide tend to gather together in aqueous solution by hydrophobic interaction, the theory attracted interest from many fields of chemistry. As we see in the micelle formation, hydrophobic interaction proceeds with an increase in enthalpy and a larger increase in entropy ($\Delta H = +0.2$–$0.5\,\mathrm{kcal\,mol^{-1}}$ and $\Delta S = $ ca. $33\,\mathrm{cal\,K^{-1}\,mol^{-1}}$). The hydrophobic interactions result in the formation of organic clusters, and hence must include some negative change in entropy. The negative entropic contribution by the cluster formation must be compensated for by the larger increase in entropy by the solvent water. In aqueous solutions, a nonpolar solute molecule is believed to be surrounded by a clathrate cage structure of water molecules, known as an "iceberg."[37] During the process of cluster formation, the organized water cage surrounding the individual organic molecule is destroyed. The water molecules are liberated from the constraining surroundings and produce an increase in entropy. More recent investigations showed that creation of the cavity to accommodate the hydrocarbon solute molecules plays a more important role than the actual structural change of water.

The inclusion power of cyclodextrin and the molecular recognition power of enzymes have been partly attributed to hydrophobic interaction. In an aqueous solution, the CH/π contiguous folded conformations or the intermolecular approach of the hydrocarbon (part of) molecules must be assisted to some extent by hydrophobic interaction.

REFERENCES

1. G. C. Pimentel and A. L. McClellan, *The Hydrogen Bond*, Freeman & Co., San Francisco, 1960.
2. R. D. Green, *Hydrogen Bonding by CH Group*, Wiley, New York, 1984.
3. E. Beckman and O. Faust, *Z. Phys. Chem.*, **89**, 235, 247 (1914).
4. A. Allerhand and P. von R. Schleyer, *J. Am. Chem. Soc.*, **85**, 1715 (1963).

5. C. J. Creswell and A. L. Allred, *J. Am. Chem. Soc.*, **85**, 1723 (1963).

6. A. Goel and C. N. R. Rao, *Spectrochim. Acta*, **27**, 2828 (1971).

7. P. L. Olympia, Jr., *Chem. Phys. Lett.*, **5**, 593 (1970).

8. R. Taylor and O. Kennard, *J. Am. Chem. Soc.*, **104**, 5063 (1982).

9. A. Doubleday, H. Higgs, T. Hummelink, B. G. Hummelink-Peters, O. Kennard, W. D. S. Motherwell, J. R. Rodgers, and D. G. Watson, *Acta Cryst. B*, **35**, 2331 (1979).

10. A. Bondi, *J. Phys. Chem.*, **68**, 441 (1964).

11. J. A. R. P. Sarma and G. R. Desiraju, *J. Chem. Soc.*, *Perkin 2*, 1987, 1195.

12. S. Pinchas, *Anal. Chem.*, **27**, 2 (1955), *ibid.*, **29**, 334 (1957); *Chem. & Ind.*, 1451 (1959).

13. S. Pinchas, *J. Phys. Chem.*, **67**, 1862 (1963).

14. W. F. Forbes, *Can. J. Chem.*, **40**, 1891 (1962).

15. H. Satonaka, K. Abe, and M. Hirota, *Bull. Chem. Soc. Jpn.*, **61**, 2031 (1988).

16. J. L. Adcock and H. Zhang, *J. Org. Chem.*, **60**, 1999 (1995).

17. P. Seiler and J. D. Dunitz, *Helv. Chim. Acta*, **72**, 1125 (1989).

18. L. P. Hammett, *Physical Organic Chemistry*, McGraw Hill, New York, 1940, p. 294.

19. L. J. Andrews, *Chem. Rev.*, **54**, 713 (1954).

20. G. Herzberg, *Infrared and Raman Spectra of Polyatomic Molecules*, van Nostrand, New York, 1945.

21. N. Sheppard, *Hydrogen Bonding*, D. Hadzi and H. W. Thompson, Eds., Pergamon Press, London, 1959, pp. 85–105.

22. M. L. Josien and G. Sourisseau, *Bull. Soc. Chim. Fr.*, 178 (1955); M. L. Josien, G. Sourisseau, and C. Castinel, *ibid.*, 1539 (1955); N. Fuson, P. Pineeau, and M. L. Josien, *ibid.*, 423 (1957); M. L. Josien and G. Sourisseau, *Hydrogen Bonding*, D. Hadzi and H. W. Thompson, Eds., Pergamon Press, London, 1959, pp. 129–137.

23. R. Ehama, A. Yokoo, M. Tsushima, T. Yuzuri, H. Suezawa, K. Sakakibara, and M. Hirota, *Bull. Chem. Soc. Jpn.*, **66**, 814 (1993).

24. N. Fuson, P. Pineau, and M. L. Josien, *J. Chim. Phys.*, **55**, 454, 464 (1958).

25. A. Engdahl and N. Bengt, *Chem. Phys. Lett.*, **113**, 49 (1985).

26. M. Oki and H. Iwamura, *Bull. Chem. Soc. Jpn.*, **32**, 955 (1959); *ibid.*, **32**, 1135 (1959); *ibid.*, **35**, 1552 (1962); M. Oki, H. Hosoya, and H. Iwamura, *ibid.*, **34**, 1391, 1395 (1961); M. Oki and H. Iwamura, *J. Am. Chem. Soc.*, **89**, 576 (1967); M. Oki, H. Iwamura, T. Onoda, and M. Iwamura, *Tetrahedron*, **24**, 1905 (1968).

27. M. Oki and K. Mutai, *Bull. Chem. Soc. Jpn.*, **33**, 784 (1960); *ibid.*, **38**, 387 (1965); *ibid.*, **39**, 809 (1966).

28. H. S. Rzepa, M. L. Webb, A. M. Z. Slawin, and D. J. Williams, *J. Chem. Soc., Chem. Commun.*, 765 (1991).

29. J. L. Atwood, F. Hamada, K. D. Robinson, G. W. Orr, and R. L. Vincent, *Nature*, **349**, 683 (1991).

30. S. Ueji, K. Nakatsu, H. Yoshioka, and K. Kinoshita, *Tetrahedron Lett.*, 1173 (1982).

31. A. D. U. Hardy and D. D. MacNicol, *J. Chem. Soc., Perkin 2*, 1140 (1976).

32. T. Cairns and G. Eglinton, *J. Chem. Soc.*, 5906 (1965).

33. S. S. Al-Juaid , K. A. Al-Nasr, C. Eaborn, and P. B. Hitchcock, *J. Chem. Soc., Chem. Commun.*, 1482 (1991).

34. B. Schweizer, J. D. Dunitz, R. A. Pfund, and G. M. R. Tombo, *Helv. Chim. Acta*, **46**, 2738 (1981).

35. P. Hobza and R. Zahradnik, *Weak Intermolecular Interactions in Chemistry and Biology*, Elsevier, Amsterdam, 1980.

36. W. Kauzmann, *Adv. Protein Chem.*, **14**, 1 (1959).

37. W. Y. Wen and J. H. Hung, *J. Phys. Chem.*, **74**, 170 (1970).

CHAPTER 5

CONFORMATIONAL CONSEQUENCES

The CH/π interaction may play a role in intramolecular as well as intermolecular interactions. Intramolecular interactions include conformational consequences, chiroptical properties of unsaturated compounds, and selectivities of certain chemical reactions. In this chapter, we examine conformations of a variety of organic compounds.

5.1. CONFORMATION OF 1,5-HEXADIENES

Saito et al. studied, using the dichroic exciton chirality method and NOE, the conformation of diethyl 4,4-bis-t-butyldimethylsiloxy-2E,6E-octadiendioate.[1]

The results point to an unequivocal conclusion that the conformation as depicted in Figure 5.1, where the enoate groups are close to

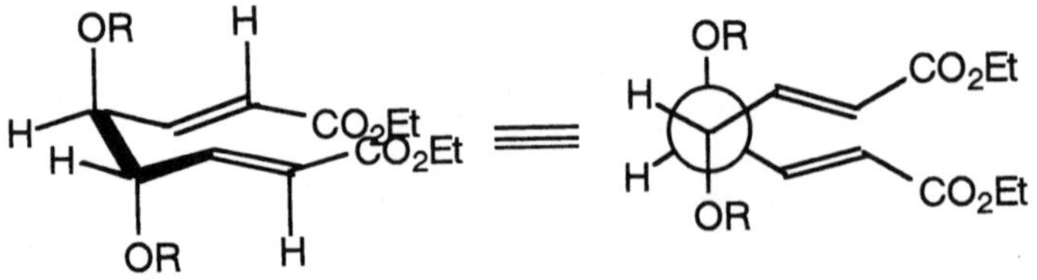

Figure 5.1. Suggested conformation for diethyl 4,4-bis-t-butyldimethyl-siloxy-2E,6E-octadiendioate.

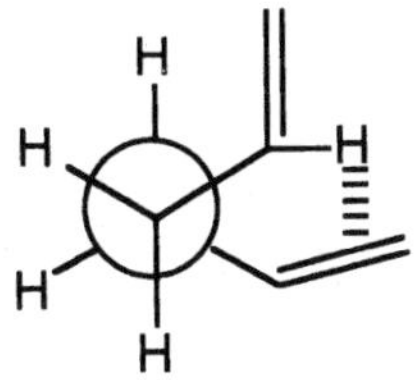

Figure 5.2. A stable conformation of 1,5-hexadiene suggested by MP2/ 6-41G* calculation.

each other, preponderates in solution. Repulsion (steric or electro- static) between the OR groups may be important, but the reason for the unusual conformation remains to be elucidated.

A rationale for this finding was presented by Gung et al. on the grounds of the CH/π interaction.[2] The conformation, as depicted in Figure 5.2, for 1,5-hexadiene has been found to be the most stable, by ab initio (MP2/6-41G*) calculations among a number of possible forms.

The conclusion was, however, incompatible with results obtained from force-field calculations. The failure of molecular mechanics in reproducing the experimental and ab initio results has been attributed

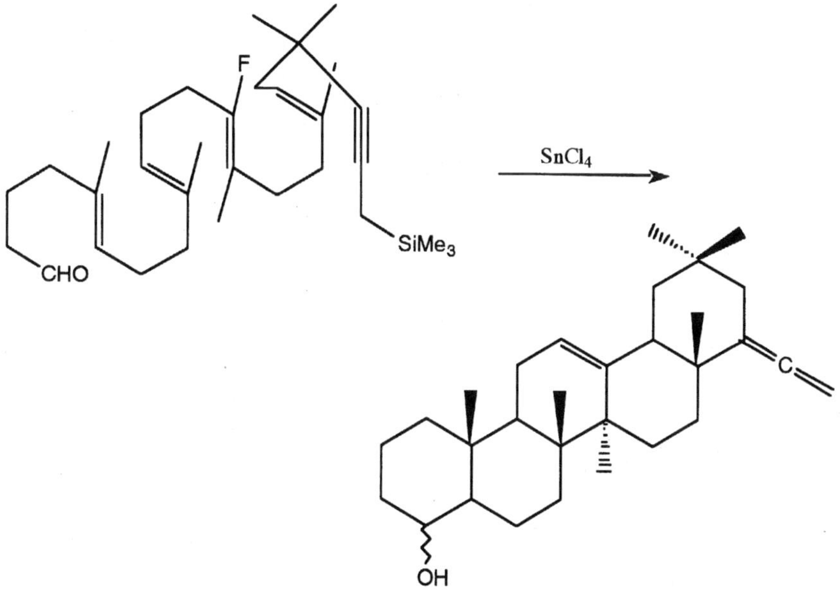

Figure 5.3. One-step cyclization reaction of a squalene derivative.

by these authors to neglecting the contribution from the attractive forces between the CH and π system. The suggestion is important, since their findings bear implications for the mechanisms of reactions like Cope rearrangements and nonenzymatic syntheses of terpenoids. To cite an example, Fish and Johnson reported that the cyclization, as illustrated in Figure 5.3, proceeded with 49% yield in a one-step reaction.[3]

Such a remarkable selectivity is hardly conceivable without invoking an attractive interaction between the relevant groups in stabilizing the geometry of the transition state, as indicated in Figure 5.1. The driving force may include a CH/π-type interaction, but this suggestion must await further clarification.

5.2. SUBSTITUENT EFFECT ON CONFORMATIONAL ENERGIES IN ATROPISOMERISM

Wilcox et al. studied the conformations of esters of dibenzodiazocines **1**.[4] Rotation about the biphenyl moiety is slow in these compounds and allows separate observation of NMR peaks for the folded and extended conformers (Fig. 5.4).

For the folded conformation, the aromatic ring in the ester part lies over another ring situated at the terminus of the molecule, orientating itself in a tilted-T position relative to the ring. The crystal structure of the *p*-nitrophenyl ester ($X = NO_2$) revealed that the two rings are in a nearly perfect T-relationship. The relevant centroid distance is short, 4.95 Å.

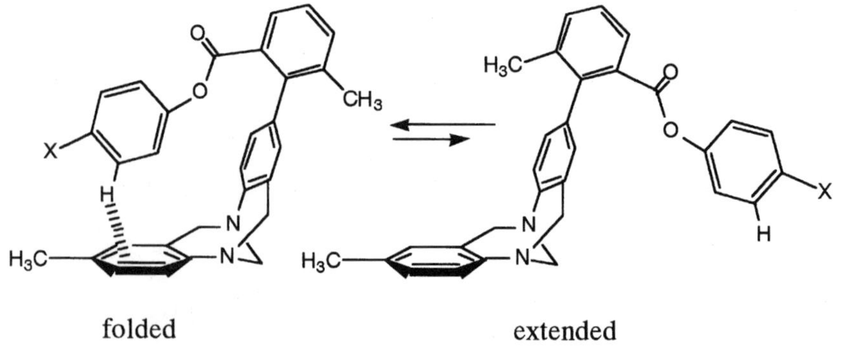

folded   extended

Figure 5.4. Conformations of dibenzodiazocine esters **1**.

TABLE 5.1. Rotameric Equilibria of Dibenzodiazocine Esters 1

Substituent	Percent Folded[a]	ΔG (kcal mol^{-1})
C_6H_5	60	0.24
p-$CH_3C_6H_4$	65	0.37
p-$CH_3OC_6H_4$	60	0.24
p-CNC_6H_4	75	0.65
p-$NO_2C_6H_4$	75	0.65
m-$CH_3C_6H_4$	50	0
3,5-diMeC$_6$H$_3$	35	-0.37

[a] In CDCl$_3$ at 298 K.

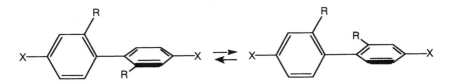

Figure 5.5. Atropisomerism of biphenyl derivatives **2** and **3**.

They examined, by ^{1}H NMR, the effect of substitution of a group (X) on the rotameric equilibria of **1**. Table 5.1 lists the results. A remarkable fact is that substitution of X from H by an electron-withdrawing group is accompanied by an increase in the ratio of folded conformers. Replacement of H by methyl group(s) at meta position(s) decreased the stability of the folded conformer. Solvent will modulate the result, if the electrostatic effect or the solvation energies are important in determining the rotameric equilibria. Changes in the solvent gave, however, negligible effect on the rotamer ratios. These findings may be accommodated in the context of the CH/π interaction.

Wolf et al. studied the effect of substituents (X) on the rotational energy barrier of 2,2'-bis(trifluoromethyl)biphenyl and 2,2'-diiso-propylbiphenyl derivatives (Fig. 5.5, R = CF$_3$ **2** and R = CHMe$_2$ **3**) by dynamic gas chromatography.[5]

The effect of para-substituent X on the rate of rotation of the chiral biphenyl depends on the aromatic system (R = CF$_3$ or CHMe$_2$)

TABLE 5.2. Activation Free Energy of the Rotation [$\Delta G^{\neq}$ (kcal mol^{-1})] of 2 and 3 Determined by Dynamic Gas Chromatography (298 K)

X	2	3
NH_2	23.8	27.6
CH_3	24.8	
H	25.6	26.7
NO_2	26.2	25.2
F	26.2	

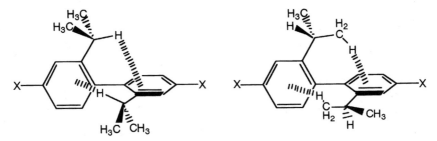

Figure 5.6. Stabilization of the ground-state structure of 2,2'-diisopropyl-biphenyl derivatives **3**.

to which they are attached. Electron-donating groups decrease the rotational energy barrier of **2** but increase that of **3**. Electron-with-drawing groups exhibited the opposite behavior (Table 5.2.)

They discussed the unusual results in terms of the CH/π interaction hypothesis. Thus, hydrogens of the isopropyl group (CH and CH_3) of **3** can participate in interactions with the aromatic part of the oppos-ing ring. The CH/π interactions illustrated in Figure 5.6 may contrib-ute to stabilize the ground-state structure of **3**. These interactions are not effective in the transition state where the two aromatic rings are almost coplanar.

Electron-donating groups will increase the π-donating property of the phenyl group and thus support the charge transfer. The reverse is true for electron-withdrawing substituents. Stabilization in the ground state of **3** will result in increase of the activation free energy of the rotation.

5.3. CONFORMATION OF VARIOUS ORGANIC COMPOUNDS

Presence of the conformers with alkyl/aryl and aryl/aryl contiguous geometry has often been noted. In this section are compiled reports which may support the CH/π interaction hypothesis. The structures and/or names of the compounds are given in Tables 5.3 and 5.4 together with the methods.

TABLE 5.3. Simple Organic Compounds

Alkyl/π-System Interactions

3-Phenyl-3,5,5-trimethylcyclohexanone (NMR)[6]

3-Aryl-1,3,5,5-tetramethylcyclohexanol (NMR)[7]

Chloroquine, an antimalarial (LIS)[8]

$C_6H_5CHCH_3CH(OH)R$,[9] $C_6H_5CHCH_3SOR$,[10] $C_6H_5CH_2CH(OH)R$[11] R = Me, Et, Pr^i, Bu^t

TABLE 5.3   *(Continued)*

$C_6H_5CHCH_3COR;$ R = H, Me, Et, Pr^i, Bu^t (LIS)[12]

1-Benzazocinone (X-ray, NOE)[13]

A benzothiopyran derivative (X-ray, NMR)[14]

α-Phellandrene (ORD)[15]

Levopimaric acid (ORD, [16] X-ray[17])

TABLE 5.3 (*Continued*)

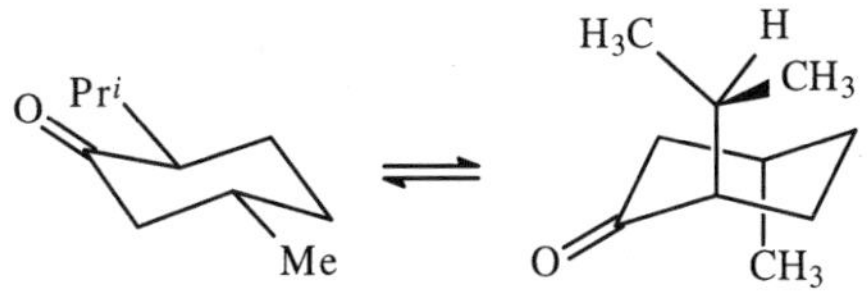

Menthone (ORD, [18] LIS[19])

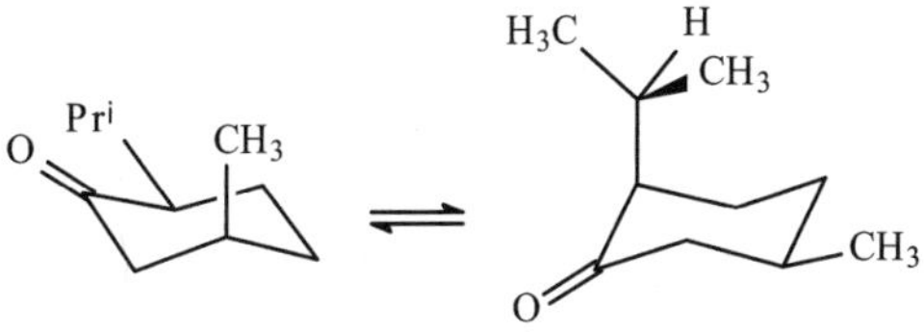

Isomenthone (ORD) [20]

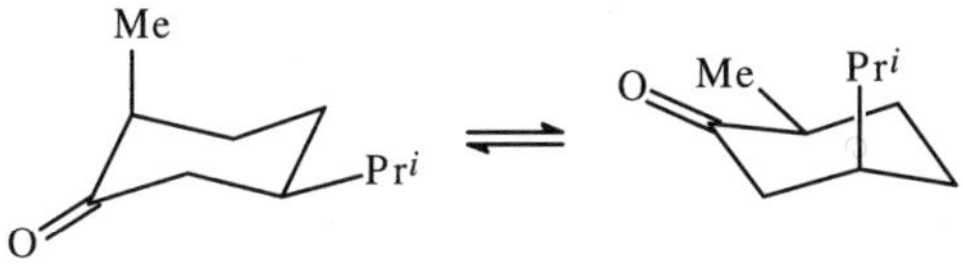

Isocarvomenthone (LIS)

Aryl/Aryl Interactions

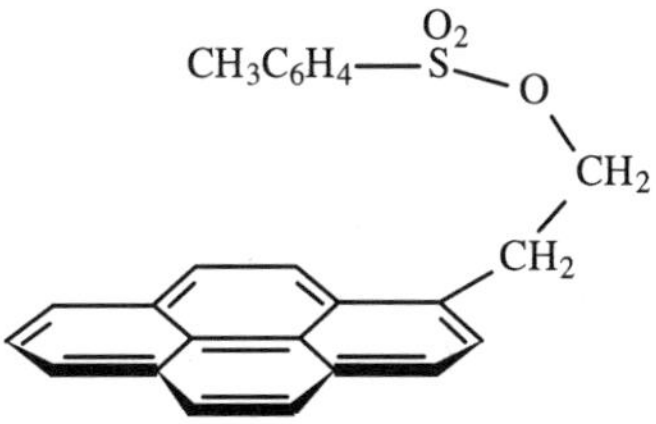

2-(1-Pyrenyl)ethyl *p*-toluenesulfonate and analogs (NMR)[21]

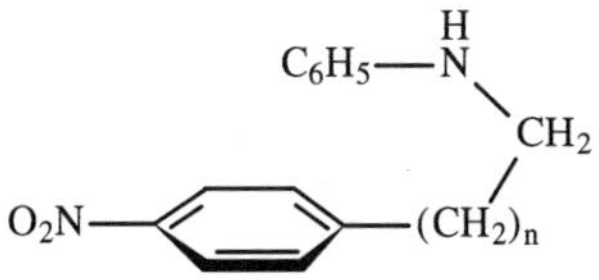

p-NO$_2$C$_6$H$_4$(CH$_2$)$_n$NHC$_6$H$_5$, n = 1–3 (UV)[22]

ArSO$_2$CH$_2$Ar, ArSOCH$_2$Ar (UV, NMR, X-ray)[23]

TABLE 5.3   (*Continued*)

Aryl-*N*-(arylsulfonylmethyl)-*N*-methylcarbamates (UV, NMR, X-ray)[24]

$C_6H_5CH(R)SOAr$; R = CH_3, C_2H_5 (NMR, LIS)[25]

$C_6H_5CHCH_3CH(OH)Ar(LIS)$[26]

Bisparaphenylene-25-crown-7 (X-ray)[27]

1,5-Dihydroxynaphtho-35-crown-9 (X-ray, NMR)[28]

Z-*N*-[1-(1-naphthyl)ethylidene]-1-phenyl-2-propylamine (NMR, X-ray)[29]

TABLE 5.3   *(Continued)*

1,4-Dihydro-4-tritylbiphenyl (X-ray, NMR)[30]

TABLE 5.4. Biochemically Relevant Molecules

Peptides and Analogs

c-L-Leu-L-Tyr (NMR,[31] X-ray[32])

Flavinyl peptides (Fluorometry)[33]
Ilamycin, a cyclic heptapeptide (NMR,[34] X-ray[35])

Somatostatin (CD,[36] NMR[37])

Gly[6]-bradykinin, Gly-Pro-Phe (NMR)[38]

Lys-Phe-Phe, Phe-Gly-Gly-L-Phe, Phe-Gly-Gly-D-Phe, phenylalanylbenzyl ester, *N*-phenyl-acetyl-phenylalanine (X-ray)[39]

TABLE 5.4 *(Continued)*

A tryptophane derivative (X-ray)[40]

D-Arg-L-Phe-NHBz, Ala-Phe-OBz, Val-Phe-OBz, Leu-Phe-OBz,
Arg-Phe-OBz (NOE)[41]

Cyclo-D-Tyr-Arg-Gly-Asp-Phe-Gly (NMR)[42]

Nucleotides and Analogs

Pyridine dinucleotides (NMR)[43]
Nicotineamide–adenine dinucleotide analogs (UV)[44]
N-Methyl-N-ethylnicotineamide-adenine dinucleotide (NMR)[45]
N-[3-(Aden-9-yl)propyl]-3-carbamoylpyridinium bromide (X-ray)[46]
Mixed ligand-metal complexes of ATP and tryptophane, etc.
(Potentiometry, NMR)[47]
[Cu(thiamine pyrophosphate)(phen)] (X-ray)[48]
[Cu(5'-AMP)(bpy)]$_2$ (X-ray)[49]

REFERENCES

1. S. Saito, O. Narahara, T. Ishikawa, M. Asahara, and T. Moriwake, *J. Org. Chem.*, **58**, 6292 (1993).

2. B. W. Gung, Z. Zhu, and R. A. Fouch, *J. Am. Chem. Soc.*, **117**, 1783 (1995).

3. P. V. Fish and W. S. Johnson, *J. Org. Chem.*, **59**, 2324 (1994). See also A. Eschenmoser, L. Ruzicka, O. Jeger, and D. Arigoni, *Helv. Chim. Acta*, **226**, 1890 (1955).

4. S. Paliwal, S. Geib, and C. S. Wilcox, *J. Am. Chem. Soc.*, **116**, 4497 (1994).

5. C. Wolf, D. H. Hochmuth, W. A. König, and C. Roussel, *Liebigs Ann.*, 357 (1996).

6. B. L. Shapiro, M. J. Gattuso, N. F. Hepfinger, R. L. Shone, and W. L. White, *Tetrahedron Lett.*, 219 (1971); B. L. Shapiro and M. M. Chrysam, III, *J. Org. Chem.*, **38**, 880 (1973).

7. B. L. Shapiro, M. D. Johnston, Jr., and M. J. Shapiro, *J. Org. Chem.*, **39**, 796 (1974).

8. N. S. Angerman, S. S. Danyluk, and T. A. Victor, *J. Am. Chem. Soc.*, **94**, 7137 (1972).

9. Y. Kodama, K. Nishihata, S. Zushi, M. Nishio, J. Uzawa, K. Sakamoto, and H. Iwamura, *Bull. Chem. Soc. Jpn.*, **52**, 2661 (1979); J. Uzawa, S. Zushi, Y. Kodama, Y. Fukuda, K. Nishihata, K. Umemura, M. Nishio, and M. Hirota, *ibid.*, **53**, 3623 (1980).

10. Y. Kodama, S. Zushi, K. Nishihata, M. Nishio, and J. Uzawa, *J. Chem. Soc., Perkin 2*, 1306 (1980).

11. S. Zushi, Y. Kodama, K. Nishihata, K. Umemura, M. Nishio, J. Uzawa, and M. Hirota, *Bull. Chem. Soc. Jpn.*, **54**, 2113 (1981).

12. S. Zushi, Y. Kodama, Y. Fukuda, K. Nishihata, K. Umemura, M. Nishio, and M. Hirota, *Bull. Chem. Soc. Jpn.*, **53**, 3631 (1980).

13. K. Oda, T. Ohnuma, Y. Ban, and K. Aoe, *J. Am. Chem. Soc.*, **106**, 5378 (1984).

14. H. Shimizu, T. Yonezawa, and T. Watanabe, *Chem. Commun.*, 1659 (1996).

15. H. Ziffer, E. Charney, and U. Weiss, *J. Am. Chem. Soc.*, **84**, 2961 (1962); G. Snatzke, E. sz Kovats, and G. Ohloff, *Tetrahedron Lett.*, 4551 (1966).

16. A. W. Burgstahler, H. Ziffer, and U. Weiss, *J. Am. Chem. Soc.*, **83**, 4660 (1961); A. W. Burgstahler, J. Gawronski, T. F. Niemann, and B. A. Feinberg, *J. Chem. Soc., Chem. Commun.*, 121 (1971).

17. U. Weiss, W. B. Whalley, and I. L. Karle, *J. Chem. Soc., Chem. Commun.*, 16 (1972).

18. V. M. Potapov, G. V. Kirushkina, and A. P. Terent'ev, *Dokl. Akad. Nauk SSSR*, **189**, 338 (1969).

19. J. D. Roberts, G. E. Hawkes, J. Husar, A. W. Roberts, and D. W. Roberts, *Tetrahedron*, **30**, 1833 (1974).

20. C. Djerassi, *Optical Rotatory Dispersion*, McGraw-Hill, New York, 1960, pp. 106, 187.

21. M. D. Bentley and M. J. S. Dewar, *Tetrahedron Lett.*, 5045 (1967). See also M. J. S. Dewar and C. C. Thompson, Jr., *Tetrahedron, Suppl.*, **7**, 97 (1966).

22. M. Oki and K. Mutai, *Tetrahedron*, **26**, 1181 (1970).

23. R. van Est-Stammer and J. B. F. N. Engberts, *Can. J. Chem.*, **51**, 1187 (1973); R. van Est-Stammer and J. B. F. N. Engberts, *Tetrahedron Lett.*, 3215 (1971); I. Tickle, J. Hess, A. Vos, and J. B. F. N. Engberts, *J. Chem. Soc., Perkin 2*, 460 (1978).

24. S. van der Werf and J. B. F. N. Engberts, *Rec. Trav. Chim.*, **90**, 663 (1971); R. van Est-Stammer and J. B. F. N. Engberts, *ibid.*, **91**, 1298 (1972); R. M. Tel and J. B. F. N. Engberts, *J. Chem. Soc., Perkin 2*, 483 (1976); R. J. J. Visser, A. Vos, and J. B. F. N. Engberts, *ibid.*, 634 (1978).

25. K. Kobayashi, T. Sugawara, and H. Iwamura, *J. Chem. Soc., Chem. Commun.*, 479 (1981); K. Kobayashi, Y. Kodama, M. Nishio, T. Sugawara, and H. Iwamura, *Bull. Chem. Soc. Jpn.*, **55**, 3560 (1982).

26. N. Kunieda, H. Endo, M. Hirota, Y. Kodama, and M. Nishio, *Bull. Chem. Soc. Jpn.*, **56**, 3110 (1983).

27. A. M. Z. Slawin, N. Spencer, J. F. Stoddart, and D. J. Williams, *J. Chem. Soc., Chem. Commun.*, 1070 (1987).

28. P. R. Ashton, E. J. T. Chrystal, J. P. Machias, K. P. Parry, A. M. Z. Slawin, N. Spencer, J. F. Stoddart, and D. J. Williams, *Tetrahedron Lett.*, 6367 (1987).

29. T. A. Hamor, W. B. Jennings, L. D. Proctor, M. S. Tolley, D. R. Boyd, and T. Mullan, *J. Chem. Soc., Perkin 2*, 25 (1990).

30. M. C. Grossel, A. K. Cheetham, D. A. O. Hope, and S. C. Weston, *J. Org. Chem.*, **58**, 6654 (1993).

31. K. D. Kopple and D. H. Marr, *J. Am. Chem. Soc.*, **89**, 6193 (1967); K. D. Kopple and M. Ohnishi, *ibid.*, **91**, 962 (1969).

32. L. E. Webb and C. Lin, *J. Am. Chem. Soc.*, **93**, 3818 (1971).

33. R. E. MacKenzie, W. Föry, and D. B. McCormick, *Biochemistry*, **8**, 1839 (1969).

34. L. W. Cary, T. Takita, and M. Ohnishi, *FEBS Lett.*, **17**, 145 (1971).

35. Y. Iitaka, H. Nakamura, K. Takada, and T. Takita, *Acta Cryst. B*, **30**, 2817 (1974).

36. L. A. Holladay and D. Puett, *Proc. Natl. Acad. Sci., USA*, **73**, 1199 (1976).

37. B. H. Arison, R. Hirschmann, W. J. Paleveda, S. F. Brady, and D. F. Veber, *Biochem. Biophys. Res. Commun.*, **100**, 1148 (1981).

38. R. E. London, J. M. Stewart, R. Williams, J. R. Cann, and N. A. Matwiyoff, *J. Am. Chem. Soc.*, **101**, 2455 (1979); M. J. O. Anteunis, F. A. M. Borremans, J. M. Stewart, and R. E. London, *ibid.*, **103**, 2187 (1981).

39. S. K. Burley, A. H. Wang, J. R. Votano, and A. Rich, *Biochemistry*, **26**, 5091 (1987).

40. T. Ishida, M. Tarui, Y. In, M. Ogiyama, M. Doi, and M. Inoue, *FEBS Lett.*, **333**, 214 (1993); Y. Shimohigashi, I. Maeda, T. Nose, K. Ikesue, H. Sakamoto, T. Ogawa, Y. Ide, M. Kawahara, T. Nezu, Y. Terada, K. Kawano, and M. Ohno, *J. Chem. Soc., Perkin 1*, 2479 (1996).

41. I. Maeda, Y. Shimohigashi, I. Nakamura, H. Sakamoto, K. Kawano, and M. Ohno, *Biochem. Biophys. Res. Commun.*, **193**, 428 (1993); H. Sakamoto, Y. Shimohigashi, I. Maeda, T. Nose, K. Nakashima, I. Nakamura, T. Ogawa, K. Kawano, and M. Ohno, *J. Molec. Recogn.*, **6**, 95 (1993).

42. R. S. McDowell, T. R. Gadick, P. L. Baker, D. J. Burdick, K. S. Chan, C. L. Quan, N. Skelton, M. Struble, E. D. Thorsett, M. Tischler, J. Y. K. Tom, T. R. Webb, and J. P. Burnier, *J. Am. Chem. Soc.*, **116**, 5069 (1994).

43. O. Jardetzky and N. C. Wade-Jardetzky, *J. Biol. Chem.*, **241**, 85 (1966); W. D. Hamill, Jr., R. J. Pugmire, and D. M. Grant, *J. Am. Chem. Soc.*, **96**, 2885 (1974).

44. N. J. Leonard, T. G. Scott, and P. C. Huang, *J. Am. Chem. Soc.*, **89**, 7137 (1967); N. J. Leonard, H. Iwamura, and J. Essinger, *Proc. Natl. Acad. Sci., USA*, **64**, 352 (1969).

45. R. H. Sarma and N. O. Kaplan, *Biochemistry*, **9**, 539 (1970); R. H. Sarma, M. Moore, and N. O. Kaplan, *ibid.*, **9**, 549 (1970); R. H. Sarma, R. J. Mynott, F. E. Hruska, and D. J. Wood, *Can. J. Chem.*, **51**, 1843 (1973).

46. P. L. Johnson, J. K. Frank, and I. C. Paul, *J. Am. Chem. Soc.*, **95**, 5377 (1973).

47. H. Sigel and C. F. Naumann, *J. Am. Chem. Soc.*, **98**, 730 (1976); P. R. Mitchell and H. Sigel, *ibid.*, **100**, 1564 (1978); P. Mitchell, B. Prijs, and H. Sigel, *Helv. Chim. Acta*, **62**, 1723 (1979). Review: H. Sigel, *Pure Appl. Chem.*, **61**, 923 (1989).

48. K. Aoki, *J. Am. Chem. Soc.*, **100**, 7016 (1978).

49. K. Aoki and H. Yamazaki, *J. Am. Chem. Soc.*, **102**, 6878 (1980).

CHAPTER 6

CHIROPTICAL PROPERTIES

6.1. CHIROPTICAL PROPERTIES OF 1,3-CYCLOHEXADIENES

In 1961, Moskowitz et al. presented an empirical rule called the diene helicity rule to explain the chiroptical properties of terpenic 1,3-diene compounds.[1]

This rule states that the sense and the amount of skewness of the chromophore in a 1,3-cisoid diene determines the sign of the Cotton effect (CE) and the rotational strength of the compound (Fig. 6.1). For example, levopimaric acid, which has a left-handed helicity, produces a negative CE ($\Delta\varepsilon - 12.2$) at the long-wavelength ($\pi–\pi^*$) transition, whereas 2,4-cholestadiene having a right-hand helix exhibited a positive CE ($\Delta\varepsilon + 12.4$).[2]

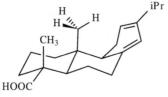

levopimaric acid ($\Delta\varepsilon - 12.2$)   2,4-cholestadiene ($\Delta\varepsilon + 12.4$)

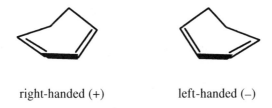

right-handed (+)   left-handed (−)

Figure 6.1. Diene helicity rule.

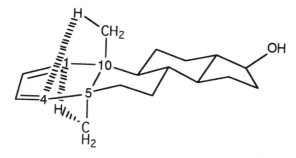

Figure 6.2. Dissymmetric perturbation of the π-orbital by methyl groups axially homoallylic to the diene system in androsta-1,3-diene.

Later, Burgstahler et al. presented evidence[3] that the contribution of an axial alkyl group allylic to the double-bond system outweighs the effect of the skewness of the diene chromophore. This is known as the concept of axial allylic chirality contribution and the chiroptical properties of cisoid dienes have since been interpreted against this background.

This phenomenon may be better accommodated in the context of the CH/π interaction.[4] Thus the important and essential condition for the enhancement of the CD amplitude is not that an alkyl group is axially allylic to the double bond, but that it is orientated suitably for CH/π interaction. In the compound illustrated in Figure 6.2, we see that the methyl groups at C^{10} and C^5 position themselves suitably to interact with C^4 and C^1, respectively. The CE amplitude of estra-1,3-diene is in fact significantly influenced by the introduction of a methyl group at C^{10} and then at C^5. The so-called axial allylic effect operates primarily through the dissymmetric perturbation of the π-orbital of the diene chromophore by virtue of the CH groups.

Tetracyclic triterpenes such as lumisterol ($\Delta\varepsilon + 14$), ergosterol (-11.4), pyrocalciferol ($+31$), isopyrocalciferol ($+25$), and 3β-hydroxycholesta-5,7-diene (-11.4) are reported to have a large CD at a wavelength corresponding to the π–π^* transition. These compounds have an angular methyl group which is ideally positioned for the CH/π interaction to occur. On the other hand, 3β-hydroxy-19-nor-cholesta-5,7-diene, which lacks the methyl group to interact with the diene system, exhibits only a small CE amplitude ($\Delta\varepsilon + 5.6$).

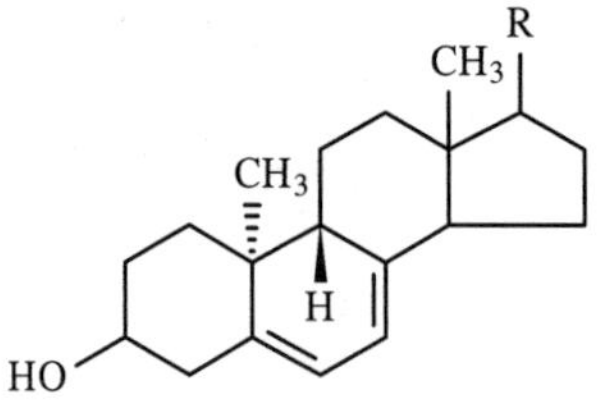

lumisterol ($\Delta\varepsilon + 14$)

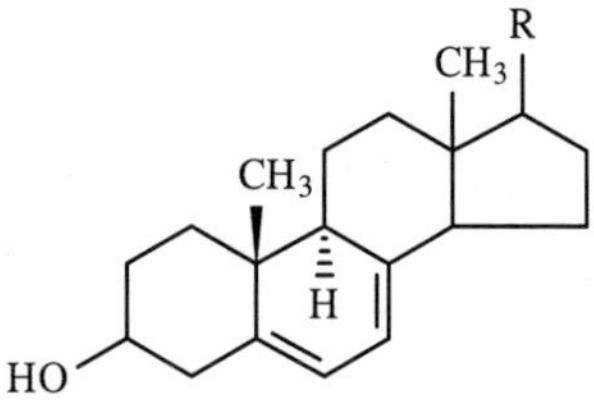

ergosterol ($\Delta\varepsilon - 11.4$)

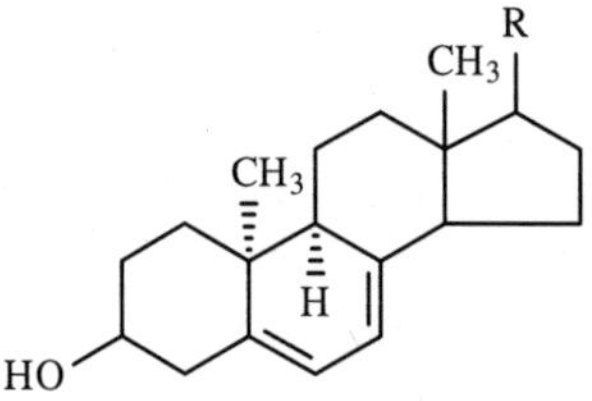

pirocalciferol ($\Delta\varepsilon + 31$)

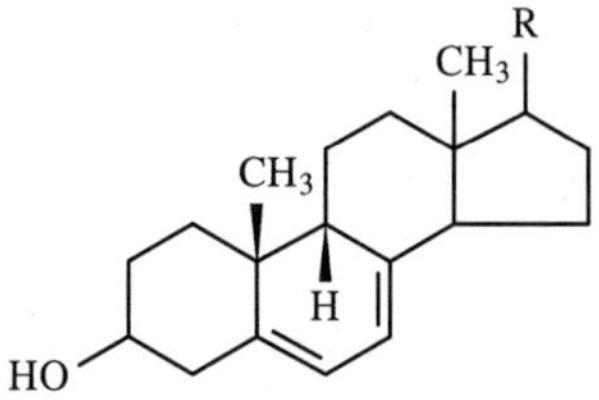

isopirocalciferol ($\Delta\varepsilon + 25$)

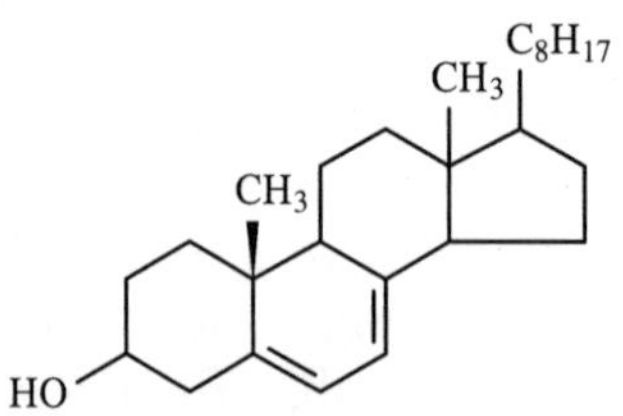

3β-hydroxycholesta-5,7-diene ($\Delta\varepsilon - 11.4$)

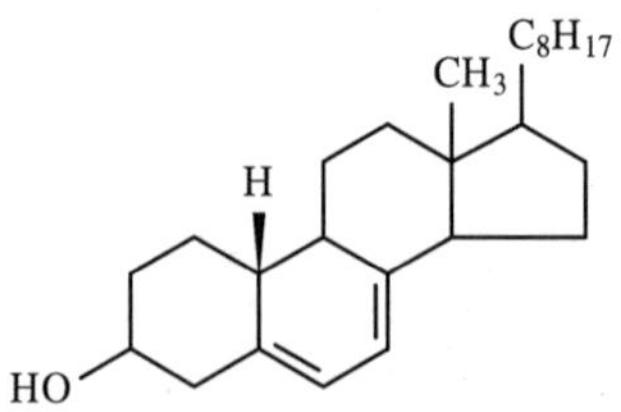

3β-hydroxy-19-norcholesta-5,7-diene ($\Delta\varepsilon + 5.6$)

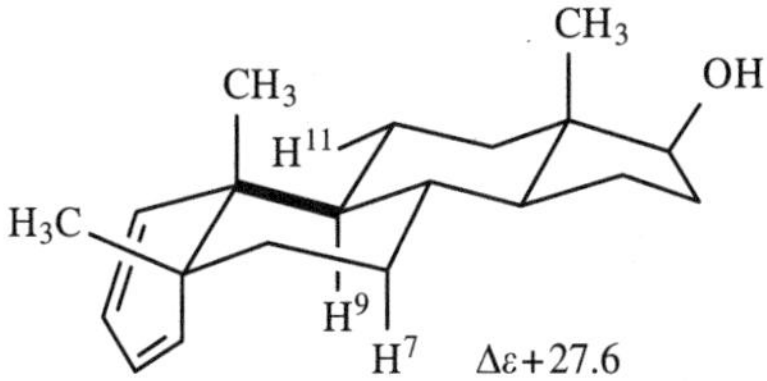

1

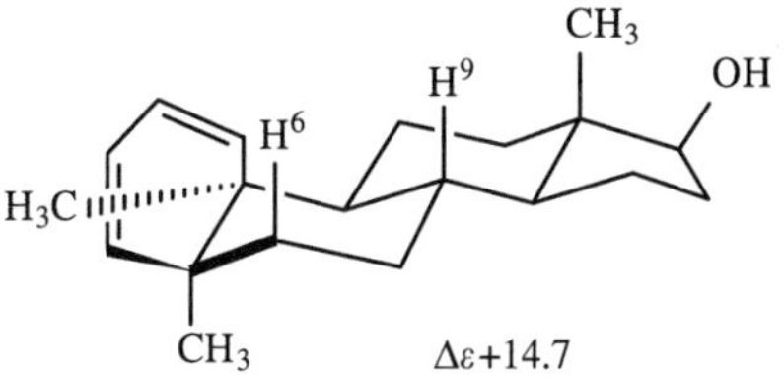

2

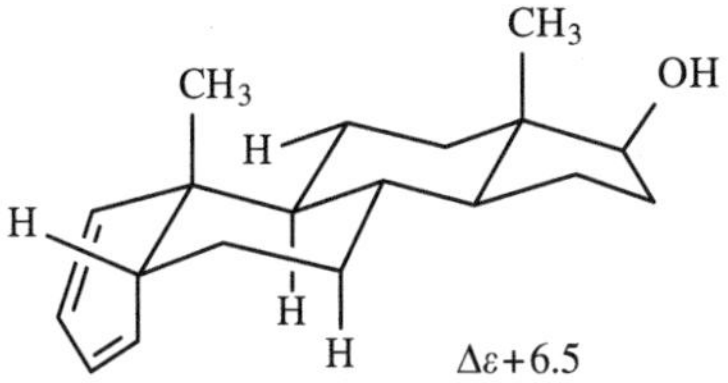

5-nor-**1**

Burgstahler et al. reported that the CE amplitude of an A/B cis compound **1** ($\Delta\varepsilon + 27.6$) is larger than that of compound **2** ($\Delta\varepsilon + 14.7$) and attributed this to the difference in polarizability of the axial alkyl group in these two compounds. Thus, in the former the "bulky" C^9 tertiary allylic axial substituent (bond boldface in structures **1** and **2**) at C^{10} brings about a stronger chirality contribution. In the latter, the "smaller" C^6 secondary allylic axial group at C^5 exerts a weaker effect.

In terms of the CH/π interaction, the explanation is simpler and clearer. We note that the number of hydrogens which can interact in a through-space manner with the diene system is three (H^7, H^9, and H^{11}) in compound **1**, whereas in **2** it is only two (H^6 and H^8). In 5-nor compound, the CE is significantly reduced ($\Delta\varepsilon + 6.5$) due to the absence of the axial methyl group at C^5 to interact with the diene chromophore.

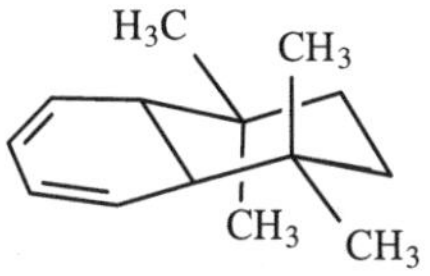

3

They found also that, in compound **3**, the presence of a methyl group at the homoallylic position exerted a significant influence on the CD spectrum.[5] The two axial methyl groups are situated at locations capable of participating in 1,5-CH/π interactions in the chair cyclohexane conformation as shown. The weak CD band with a positive CE ($\Delta\varepsilon_{267}$ + 0.8 at a 293 K) in fact shows the inversion of sign and an increase in the CE amplitude on cooling ($\Delta\varepsilon_{249}$ − 3.5 at 104 K). This indicates that, at ordinary temperatures, a significant fraction of **3** exists within the cyclohexane ring in a flexible boat conformation. At low temperatures, the population of the chair form increases sufficiently, as do the chirality contributions (negative in sign) of the two axial homoallylic methyl groups to the CE.

4 ($\Delta\varepsilon_{259}$ − 5.5)   5 ($\Delta\varepsilon_{260}$ − 1.4)

6β-Methyl-5α-cholesta-1,3-diene **4** has a CE appreciably greater than that of 6-α-Me-5α-cholesta-1,3-diene **5**. This is reasonable because **4** bears two axial methyl groups which can interact with the π-system, whereas **5** has only one axial methyl.

6.2. CHIROPTICAL PROPERTIES OF OLEFINIC COMPOUNDS

Circular dichroism of exomethylene steroids[6] show significant enhancement when an axial alkyl group is present homoallylic to the double bond. Table 6.1 lists data reported by Hudec and Kirk.[7] In every case where a significant change in the CD amplitude is shown, 1,5-CH/π interaction is possible. This is illustrated for the representative cases of 2-methylene-5α-androstane, 4-methylene-5α-androstane, 6-methylene-5α-androstane, and an 8-methylene compound, deoxyonocerine (Fig. 6.3). In the latter case, the effect seems to be duplicated, since this compound has two such interactions in a molecule.

In interpreting the structural features of olefinic compounds in relation to the CD data, however, one should keep in mind the difficulties in assigning the nature of the transitions (π–π^*, n-3s, or π–3py) which are relevant to this problem.[8]

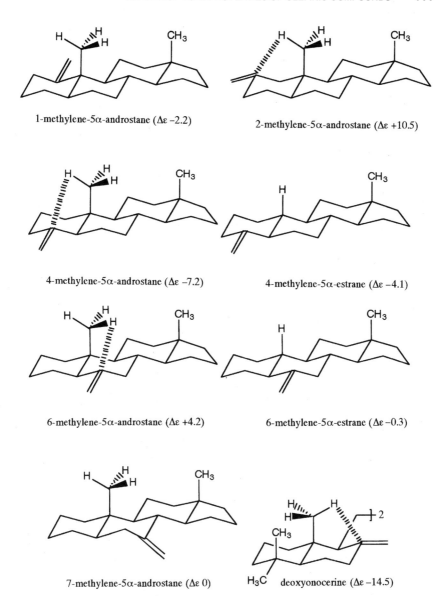

1-methylene-5α-androstane (Δε −2.2)

2-methylene-5α-androstane (Δε +10.5)

4-methylene-5α-androstane (Δε −7.2)

4-methylene-5α-estrane (Δε −4.1)

6-methylene-5α-androstane (Δε +4.2)

6-methylene-5α-estrane (Δε −0.3)

7-methylene-5α-androstane (Δε 0)

deoxyonocerine (Δε −14.5)

Figure 6.3.

TABLE 6.1. CD Spectral Data of Exomethylene Steroids (Hexane)

Compound	$\Delta\varepsilon$	λ (nm)
1-Methylene-5α-androstane	− 2.2	199
2-Methylene-5α-androstane	+ 10.5	197
3-Methylene-5α-androstane	+ 6.4	193
3-Methylene-5β-androstane	− 3.0	199
4-Methylene-5α-estrane	− 4.1	199
4-Methylene-5α-androstane	− 7.2	200
6-Methylene-5α-estrane	− 0.3	205
6-Methylene-5α-androstane	+ 4.2	197
6-Methylene-5β,25R-spirostan-3β-ol	+ 9.0	198
7-Methylene-5α-androstane	0^a	
7-Methylene-5α-cholestane	− 1.0 sh^b	200
Deoxyonocerine (8-methylene)	− 14.5	202
16-Methylene-5α-androstane	− 7.9	193
17-Methylene-5α-androstane	+ 3.8	193

^{a} J. K. Gawronski and M. A. Kielczewski, *Tetrahedron Lett.*, 2493 (1971).
^{b} sh, shoulder.

6.3. CHIROPTICAL PROPERTIES OF CYCLIC KETONES

Kirk reported that a significant effect on the shorter wavelength transition (ca. 190 nm) in various steroidal ketones and decalones is brought about by the introduction of an axial alkyl group β to the carbonyl function (equivalent to the homoallylic axial methyl in the C=C double bond system).[9] Table 6.2 lists data from their paper.

Most of the data included in Table 6.2 seems to be related to the through-space chiral contribution of the axial alkyl group positioned β to the carbonyl chromophore. 5α-Cholestan-2-one gives a positive CE ($\Delta\varepsilon$ + 4.9) dominated by a contribution from the β-axial methyl group, while CE in the 19-nor compound is negligibly small. 3β-Acetoxy-D-homo-5α-androstan-17-one, in which the relative disposition of Me with respect to C=O is approximately enantiomeric to 5α-cholestan-2-one, gives a negative CE ($\Delta\varepsilon$ − 5.0).

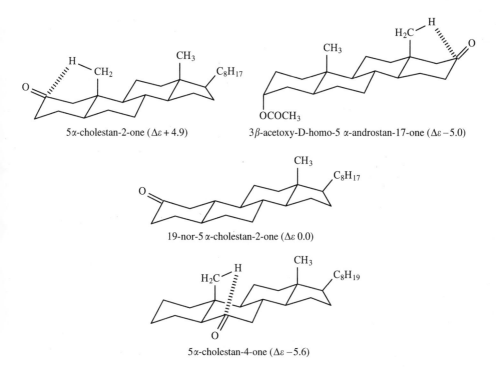

5α-cholestan-2-one (Δε + 4.9)

3β-acetoxy-D-homo-5 α-androstan-17-one (Δε −5.0)

19-nor-5 α-cholestan-2-one (Δε 0.0)

5α-cholestan-4-one (Δε −5.6)

5α-Estran-6-one gives a weak CD absorption with a positive sign, whereas the CE amplitudes of 5α-androstan-6-one, cholestan-6-one, and pregnan-6-one are significantly enhanced. 5α-Cholestan-4-one gives rise to a CD spectrum at 194 nm which is approximately the mirror image of that of the 6-oxo steroids.

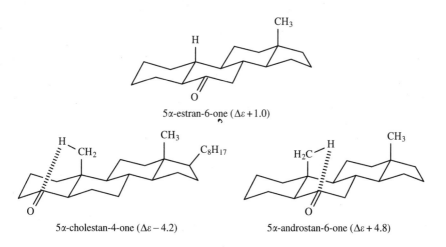

5α-estran-6-one (Δε +1.0)

5α-cholestan-4-one (Δε − 4.2)

5α-androstan-6-one (Δε + 4.8)

TABLE 6.2. Short-Wavelength CD Data of Cyclic Ketones (Hexane)

Compound	$\Delta\varepsilon$	λ (nm)
5α-Cholestan-1-one	+ 6.8	189
5α-Cholestan-2-one	+ 4.9	194
19-Nor-5α-cholestan-2-one	0.0	
5α-Cholestan-3-one	− 0.6	192
5α-Cholestan-4-one	− 5.6	191
5α-Estran-4-one	− 4.7	192
5α-Estran-6-one	+ 1.0	195
	− 3.7	186
5α-Cholestan-6-one	+ 5.1	194
5α-Androstan-6-one	+ 4.8	192
5α-Pregnan-6-one	+ 5.1	194
5α-Androstan-7-one	− 0.1	188
5α-Androstan-11-one	+ 4.5	191
5α-Androstan-12-one	+ 3.0	185
5α-Pregnan-12-one	+ 9.3	188
5α-Androstan-15-one	+ 1.8	195
3β-Acetoxy-5α-androstan-16-one	+ 3.5	193
5α-Androstan-17-one	− 8.1	193
5β-Androstan-4-one	+ 16	188
7S,9R,10S-10-Methyl-7-isopropyl-cis-1-decalone	− 12.7	185
	− 10 sh[a]	190
(2R)-2-Ethylcyclohexanone	− 1.1	191
(3R)-3-Methylcyclohexanone	+ 1.0	185
(3R)-3-Ethylcyclohexanone	+ 0.4	191
(3S)-3-Isopropylcyclohexanone	0.0	

[a] sh, shoulder.

The geometrical disposition of the axial 10-methyl group with respect to the carbonyl chromophore is enantiomeric in the 4-oxo and 6-oxo compounds.

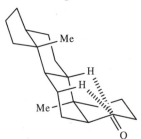

5β-androstan-4-one, $\Delta\varepsilon$ + 16

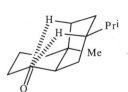

7S,9R,10S-10-methyl-7-isopropyl-cis-1-decalone, $\Delta\varepsilon$ − 12.7

5β-Androstan-4-one ($\Delta\varepsilon + 16$ at 188 nm) and $7S,9R,10S$-10-methyl-7-isopropyl-cis-1-decalone (-12.7 at 185 nm) show exceptionally large CEs. In these compounds there are two CH bonds which are orientated suitably for the through-space interaction to take place. 3-Methylcyclohexanone gives a CE with $\Delta\varepsilon + 1.0$ at 185 nm, whereas the 3-ethyl homolog shows a smaller CE ($\Delta\varepsilon + 0.4$). The CE amplitude of the isopropyl homolog is nearly zero. This is compatible with the smaller 3-alkyl ketone effect reported for 3-isopropylcyclohexanone. The discussions raised above, however, are open to further suggestions and examinations. A large $\Delta\varepsilon$ (-5.6 at 191 nm) observed for 5α-estran-4-one, for instance, cannot be explained. The nature of the relevant transitions ($n - \sigma^*$, $\pi-\pi^*$, or n-3s) still remains unclear.

Levene and Rothen studied the rotatory dispersion of a series of optically active aldehydes.[10] They observed that a positive CE was exhibited when the aldehyde function is next to the asymmetric center ($n = 0$), while separation by one carbon atom ($n = 1$) resulted in a negative CE. This was confirmed later by Djerassi and Geller[11] and similar results were also encountered with a homologous series of methyl ketones.

$$(CH_2)_nCHO \qquad\qquad (CH_2)_nCOCH_3$$

$$H_3C \!-\! C \!-\! H \qquad\qquad H_3C \!-\! C \!-\! H$$

$$C_2H_5 \qquad\qquad\qquad C_2H_5$$

Separation by more than two carbon atoms ($n > 2$) afforded a dispersion curve that is substantially more flattened compared to those of the shorter ones ($n = 0$ or 1). Also interesting is the reversal in sign of the CE, reported by Djerassi and Krakower,[12] for a homologous series of (R)-3-alkylcycloalkanones. The sign of CE is positive with 3-methylcyclopentanone and 3-methylcyclohexanone but is negative with 3-methylcycloheptanone and larger ring homolog. These phenomena may be correlated with the predominance of a certain conformation, in which the spatial situation of the carbonyl chromophore with respect to the alkyl groups is significantly different in view of the through-space orbital interaction. This explanation remains only a possibility and is open to further suggestions.

REFERENCES

1. A. Moskowitz, E. Charney, U. Weiss, and H. Ziffer, *J. Am. Chem. Soc.*, **83**, 4661 (1961).

2. A. W. Burgstahler, H. Ziffer, and U. Weiss, *J. Am. Chem. Soc.*, **83**, 4660 (1961).

3. A. W. Burgstahler, L. O. Weigel, and J. K. Gawronski, *J. Am. Chem. Soc.*, **98**, 3015 (1976).

4. S. Zushi, Y. Kodama, Y. Fukuda, K. Nishihata, M. Nishio, M. Hirota, and J. Uzawa, *Bull. Chem. Soc. Jpn.*, **54**, 2113 (1981); S. Araki, T. Seki, K. Sakakibara, M. Hirota, M. Nishio, and Y. Kodama, *Tetrahedron: Asym.*, **4**, 555 (1993).

5. A. W. Burgstahler, G. Wahl, N. Dang, M. E. Sanders, and A. Nemirovsky, *J. Am. Chem. Soc.*, **104**, 6873 (1982).

6. D. N. Kirk and W. Klyne, *J. Chem. Soc., Perkin 1*, 1076 (1974); D. N. Kirk, *ibid.*, 787 (1980); *Tetrahedron*, **42**, 777 (1986).

7. J. Hudec and D. N. Kirk, *Tetrahedron*, **32**, 2475 (1976).

8. M. Fetizon and I. Hanna, *Chem. Commun.*, 462 (1970).

9. D. N. Kirk, *J. Chem. Soc., Perkin 1*, 787 (1980).

10. P. A. Levene and A. Rothen, *J. Chem. Phys.*, **4**, 48 (1936).

11. C. Djerassi and L. E. Geller, *J. Am. Chem. Soc.*, **81**, 2789 (1959).

12. C. Djerassi and W. Krakower, *J. Am. Chem. Soc.*, **81**, 237 (1959).

CHAPTER 7

SELECTIVITY IN ORGANIC REACTIONS

Selective phenomena occur in intramolecular as well as intermolecular interactions. The former includes diastereoface-differentiating and remote functionalization reactions, while the latter includes enantio-face-differentiating and cyclization reactions. Topics that possibly implicate the CH/π interaction are compiled in this chapter.

7.1. DIASTEREOFACE-DIFFERENTIATING REACTIONS

In diastereoface-differentiating reactions such as the Prelog's system, the extent of the asymmetric synthesis has been shown to be greater for benzoyl formate (bearing a phenyl group) than for pyruvate (Me instead of Ph) (Fig. 7.1).[1]

A similar result is obtained in the case of 1-methylheptyl ester versus 1-phenylethyl ester, the former giving rise to the higher optical yield (Fig. 7.2). The results are comprehensible if an attractive interaction is operating between the alkyl and the aromatic moiety in the ground-state conformation of the substrates. Thus, in the preferred transition state, one of the two diastereofaces is preferentially attacked by the reagent.

Corey et al.[2] reported that the stereoselective reduction of a prosta-glandin precursor was better accomplished when they used esters with an aromatic substituent as R (Table 7.1).

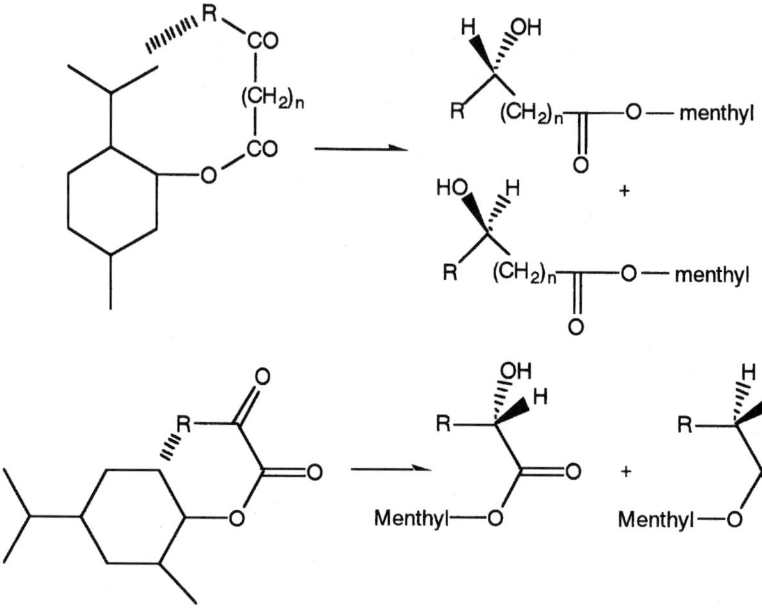

Figure 7.1. Diastereoface-differentiating reactions in the Prelog's system.

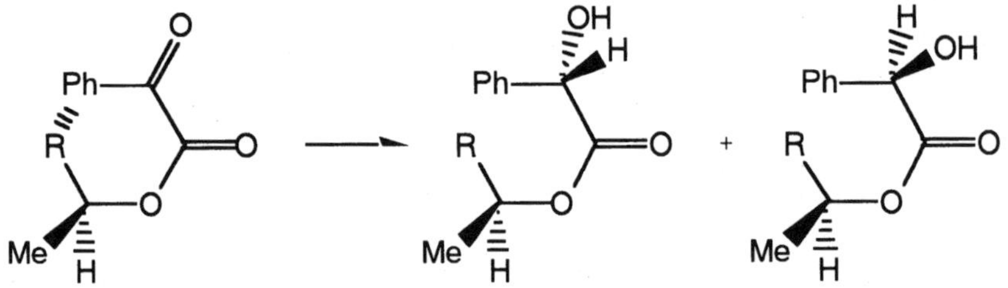

Figure 7.2. Diastereoface-differentiating reactions in the Prelog's system.

The results are reasonable because we have an aliphatic (C_5H_{11}) group at the other terminus of the molecule. A folded conformation of the substrate will prevent the reagent from approaching the other side of the carbonyl group, thus giving rise to the preferential formation of the (S)-isomer (Fig. 7.3).

Effects of changing R group on the stereochemical outcome were studied for the oxidation of a series of sulfides, $C_6H_5CH(CH_3)\text{-S-R}$ with peroxyacetic acid (Fig. 7.4).[3]

**TABLE 7.1. Stereoselective Reduction of
a Prostaglandin Precursor**

R	$(S)/(R)$ Ratio
CH_3	50/50
$CH_3(CH_2)_7$	60/40
$p\text{-}C_6H_5C_6H_4$	82/18
$p\text{-}n\text{-}C_5H_{11}C_6H_4$	82/18
C_6H_5NH	89/11
$p\text{-}C_6H_5C_6H_4NH$	92/8

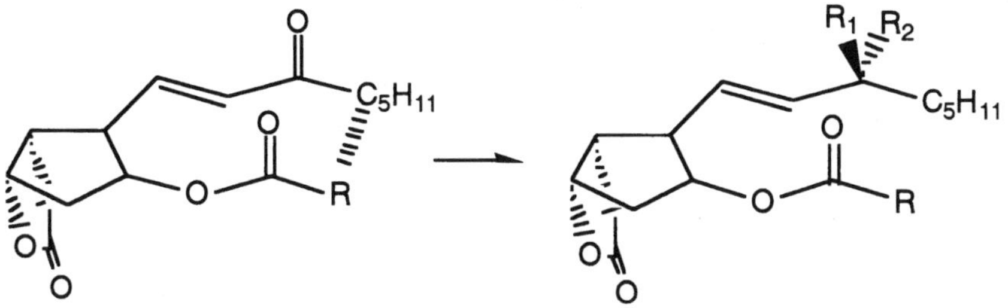

Figure 7.3 Suggested conformation of a prostaglandin precursor giving rise to preferential formation of the (S)-isomer.

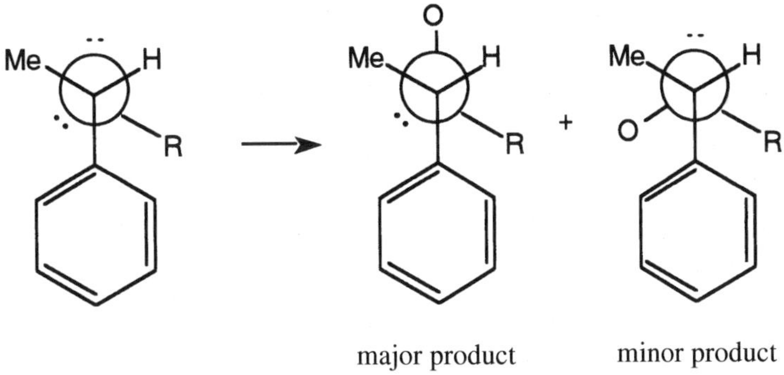

major product   minor product

Figure 7.4. Oxidation of sulfides to diastereomeric sulfoxides.

The stereoselectivity of the diastereoface-differentiating reaction varied from 3.1 for R = Me, 3.2 for Et, 3.5 for Pr^i, and to 49 for Bu^t. The results can be reasonably explained only if the conformation of the sulfides in the reactant-like transition states are synclinal with respect to the alkyl and phenyl groups.

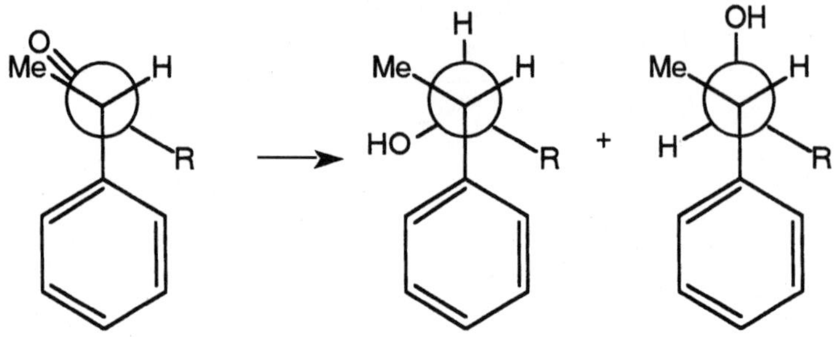

major product   minor product

Figure 7.5. LiAlH$_4$ reduction of chiral ketones to diastereomeric alcohols.

By analogy, the trends in selectivities observed in the product ratios of the LiAlH$_4$ reductions of chiral ketones, reported by Felkin et al. (Fig. 7.5; 2.8. for R = Me, 3.2. for Et, 5.0 for Pr^i, and 49 for Bu^t),[4] may also be understood. The results are reproduced moderately well by force-field calculations.[5]

Yokowo et al.[6] studied the optical activation of 2-phenylpropional-dehyde and 3-(p-cumyl)-2-methylpropionaldehyde through enamaine, employing (S)-2-isopropyl-1-methylpiperazine and (S)-3-isopropyl-1-methylpiperazine as the secondary amine components of the enamine.

The stereochemical outcome of the reaction was interpreted in view of the occurrence of the CH/π interaction between the isopropyl and phenyl groups (Fig. 7.6).

To account for the unusually high selectivities observed in the osmylation of a series of bis-allylic compounds (Fig. 7.7: R = CO$_2$Et, CH$_2$OAc, CH$_2$OBz, TBS = t-butyldimethylsilyl), Saito et al. proposed a folded conformation involved in the reactant-like ground-state.[7]

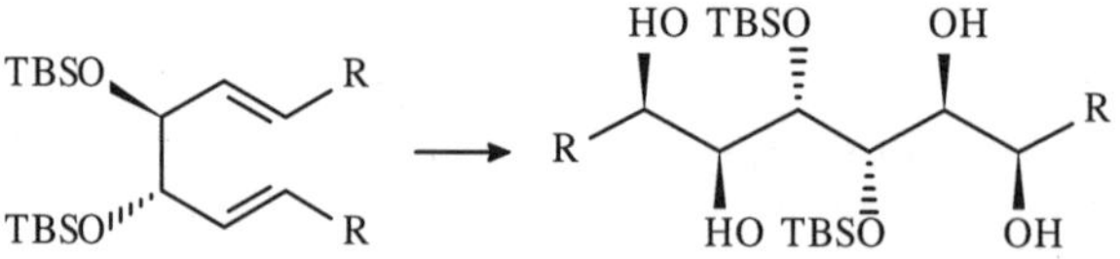

Thus diethyl (E,E)-, (Z,Z)-, and (E,Z)-4,4-bis(t-butyldimethyl-siloxy)octadiendioates (R = CO$_2$Et) give rise to 2,3,6,7-tetrahyd-roxylated products **1a**, **1b**, and **1c**, respectively, mostly as single

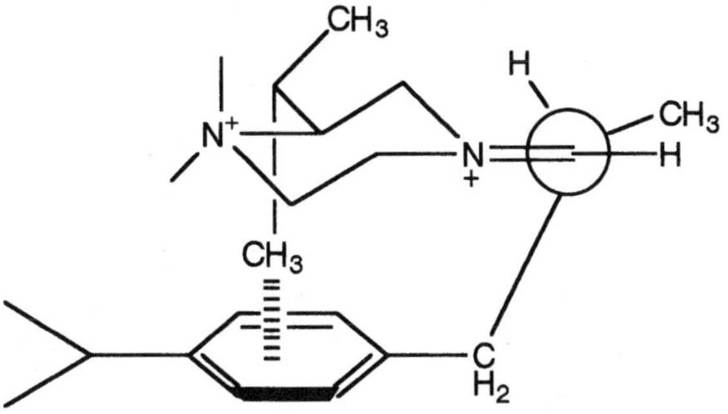

Figure 7.6. Proposed transition state for optical activation of 3-(p-cumyl)-2-methylpropionaldehyde via enamine.

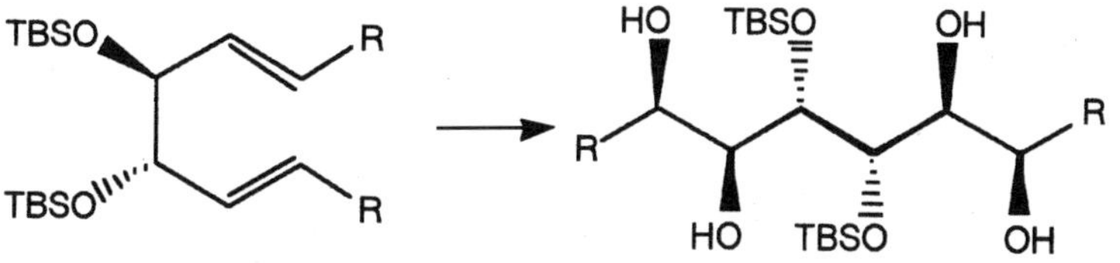

E,E

1a

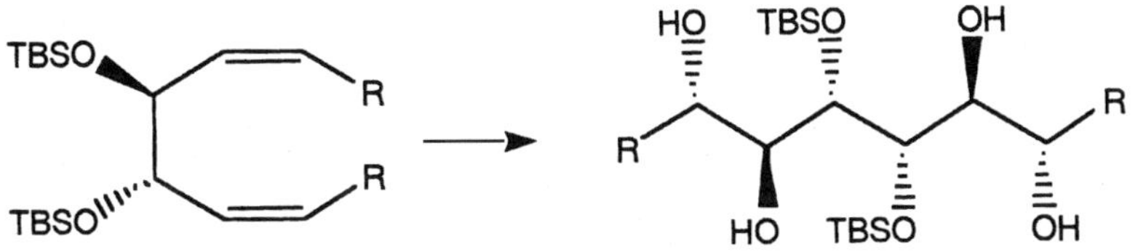

Z,Z

1b

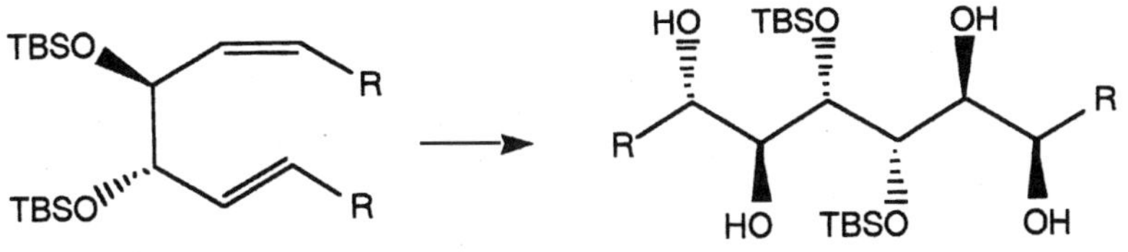

E,Z

1c

Figure 7.7. Diethyl (E,E)-, (Z,Z)-, and (E,Z)-4,4-bis(t-butyldimethylsiloxy)-octadiendioate gives rise to tetrahydroxylated products as single isomers.

isomers. This implies that the two C=C double bonds mutually shield two diastereofaces in a topological sense, thereby leaving only the other two diastereofaces available for the approach of the reagent.

7.2. REMOTE FUNCTIONALIZATION REACTIONS

In remote functionalization reactions reported by Breslow, high selectivity was always achieved whenever an aromatic group was incorporated within the reacting molecule.[8]

Thus, selective functionalization at C^{14} was achieved by applying the principle of remote oxidation of cholesterol. The transition state for the intramolecular attack may be stabilized by the attractive interaction between the benzene π-system and the steroidal (CH) part (Fig. 7.8).

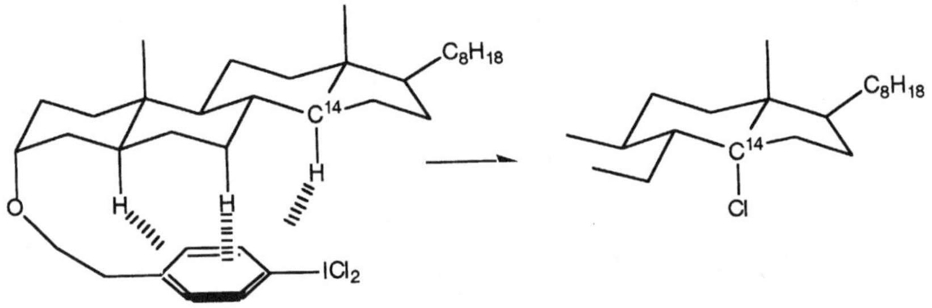

Figure 7.8. Remote chlorination of cholesterol.

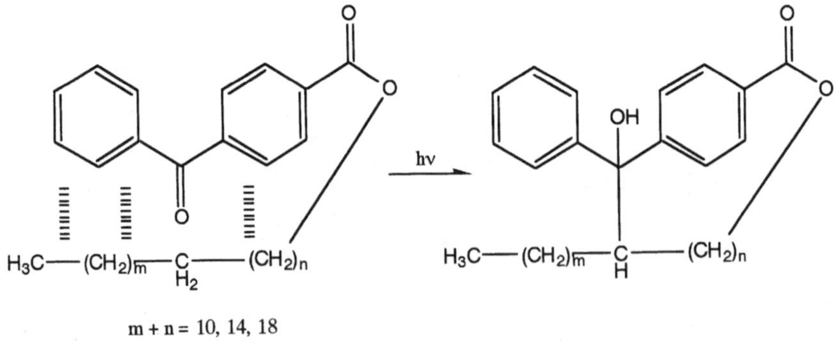

m + n = 10, 14, 18

Figure 7.9. Remote functionalization of long-chain aliphatic groups.

Breslow invoked the involvement of an extensively coiled chain to account for the selectivity obtained in the remote functionalization of long-chain aliphatic groups (Fig. 7.9).[9]

In the template-directed epoxidation of farnesol and geranylgeraniol, Breslow and Maresca found that the terpene chain is extensively coiled and folded.[10] In every case there is a preference side for attack at the end of the chain, even in cases where extensive folding must be present (Fig. 7.10). The results of Breslow may be interpreted as a consequence of the attractive interaction between the relevant groups.

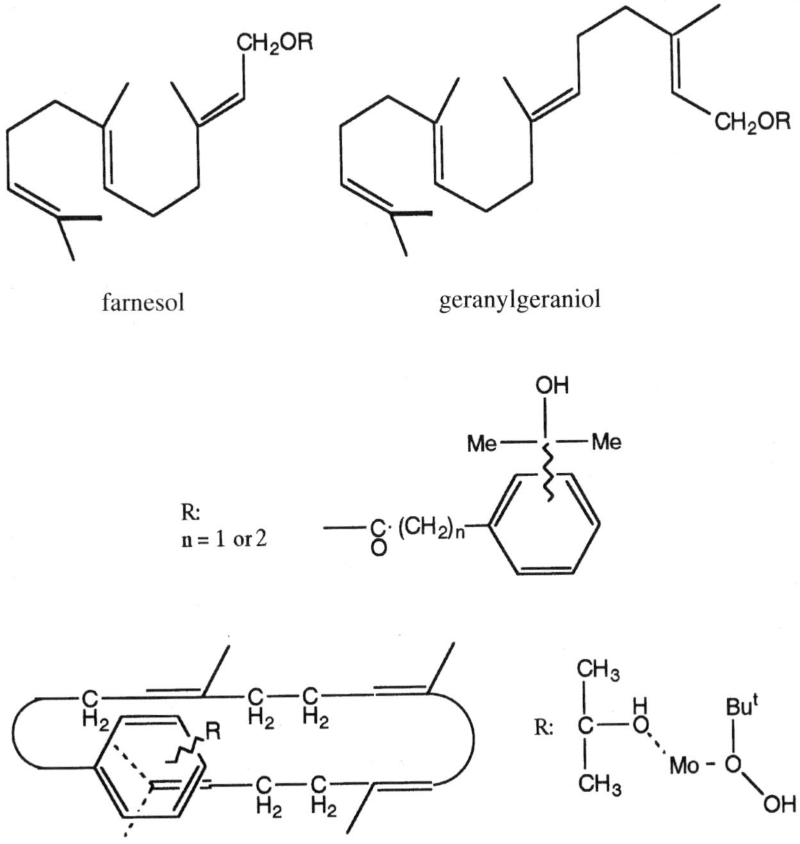

Figure 7.10. Suggested conformations of farnesol and geranylgeraniol.

7.3. ENANTIOFACE-DIFFERENTIATING REACTIONS

Tables 7.2 and 7.3 are extracted from the work of Mosher.[11] They studied the enantioface-differentiating reaction of ketones with chiral Grignard reagents prepared from (+)-1-chloro-2-methylbutane and (+)-1-chloro-2-phenylbutane. Table 7.2 gives the results obtained by reduction of $R_S COR_L$ 2 with a Grignard reagent from (+)-1-chloro-2-methylbutane 3. Table 7.3 gives the results obtained by reduction of alkyl phenyl ketones 4 with a Grignard reagent from (+)-1-chloro-2-phenylbutane 5. The chiral reduction gave rise to preferential formation of alcohols having the (S)-configuration, with a few exceptions for ketones bearing a t-butyl group.

Note that the extent of asymmetric synthesis is greater in cases where a phenyl group is incorporated in both the ketone 2 or 4 and the Grignard reagent 3 or 5. This is reasonable if we assume the CH/π interaction operating between the alkyl and the phenyl group at two

TABLE 7.2. Optical Yields (% ee) of the Reduction of Ketones 2 with Grignard Reagent from (+)-1-Chloro-2-methylbutane 3

	R_L		
R_S	Bu^t	Hex^c	Ph
Me	13	4	4
Et	11	9	6
Bu^i	6	16	10
Pr^i	0	2	24
Bu^t	—	2^a	16

^{a} (R)-enantiomer was obtained in excess.

2   3   (S)

TABLE 7.3. Optical Yields (% ee) of the Reduction of Phenyl Alkyl Ketones 4 with Grignard Reagent from (+)-1-Chloro-2-phenylbutane 5

R	R'		
	Me	Et	Pr^i
Me	38	47	—
Et	38	52	66
Bu^i	—	53	—
Pr^i	59	82	80
Bu^t	22^a	16	91

^{a} (R)-enantiomer was obtained in excess.

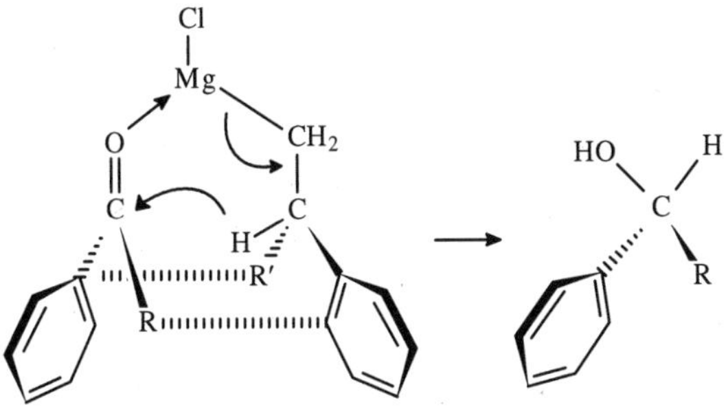

points. The attractive interaction is anticipated to increase with an increase in the number of CH groups suitably positioned with respect to Ph (Table 7.3).

Capillon and Guetté[12] studied the effects of substituents in enantioface-differentiating reductions of phenyl alkyl ketones with chiral Grignard reagents. They found that the optical yield (the preferential formation of S-enantiomer) decreased by introduction of an electron-withdrawing group on the aromatic ring of the ketones or Grignard reagents (Table 7.4).

This is reasonable because the CH/π interaction will decrease on replacement of the substituent H by CF$_3$ or Cl (Fig. 7.11). The inverse should have been true for compounds with an electron-donating substituent; this, however, is not very clear in Table 7.4.

TABLE 7.4. Optical Yields (% ee) of the Reduction of Alkyl Phenyl Ketones with Chiral Grignard Reagents

R	X/Y	OCH_3	H	CF_3
C_2H_5	OCH_3	51	51	
C_2H_5	CH_3	54	52	10
C_2H_5	H	57	50	22
$CH(CH_3)_2$	H	84	81	58
$C(CH_3)_3$	H	16	16	-27^a
C_2H_5	Cl	36	43	—
C_2H_5	CF_3	22	22	10

^{a} (R)-enantiomer was obtained in excess.

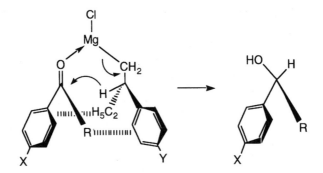

Figure 7.11. Transition state (suggested) of enantioface-differentiating reductions of phenyl alkyl ketones with chiral Grignard reagents.

Chérest and Prudent studied the stereochemistry in the hydride reduction of a series of ketones L-CHMe-CO-R (L = Ph and cyclohexyl; R = Me, Et, Pr^i, and Bu^t); the results are consistent with the presence of a methy/phenyl attractive interaction in the transition state leading to the preferred product.[13]

7.4. COUPLING REACTIONS

Kobuke et al. studied the stereochemistry of the Diels–Alder reactions of cyclopentadiene with a series of dienophiles, $CH_2=C(CH_3)X$ (Fig. 7.12).[14]

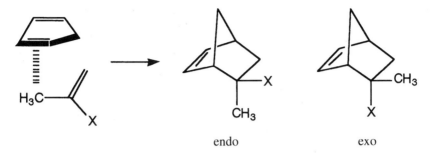

endo   exo

Figure 7.12. Diels–Alder reactions of cyclopentadiene with dienophiles, $CH_2=C(CH_3)X$. X = CN, $COCH_3$, $COOCH_3$, CHO, COOH.

They found an appreciable endo-orientating tendency for the methyl group as opposed to the polar unsaturated group, X. The result was attributed to the presence of an attractive force of the methyl group which orientates itself to stabilize the transition state leading to the endo product. Another explanation based on the repulsive interaction between electronegative CN groups and the π-group is also possible.

Closs and Moss[15] studied the effect of alkyl substitution on the stereochemical outcome in addition reactions of aryl carbenes to a series of olefins, $CH_2=CHR$ (Fig. 7.13). The syn/anti preference in the cyclopropane formation was shown to decrease as R became larger; from methyl (R = Me, 3.1), to ethyl (2.1), isopropyl (1.4), and t-butyl (0.45).

One interpretation is that the number of CH bonds involved effectively in the CH/π interaction in the transition state decreased upon successive methylation of R (CH_3 to CH_2Me, $CHMe_2$, and CMe_3). In support of this, introduction of a second methyl group to the olefin (isobutene) resulted in a significant increase of the reaction rate.[16] Syn/anti product ratio increased by substituting X from H to Me and then to MeO. This is reasonable in view of the presence of the CH/π interaction.

Yamato et al.[17] studied the coupling of 1,3-bis(bromomethyl)-7-t-butylpyrene with a series of 1,3-bis(mercaptomethyl)benzenes to give 2,11-dithia[3]metacyclo(1,3)pyrenophane derivatives. The stereochemical outcome of the reaction was found to be dependent on the nature of the substituents R^1 and R^2. The result is summarized in Table 7.5.

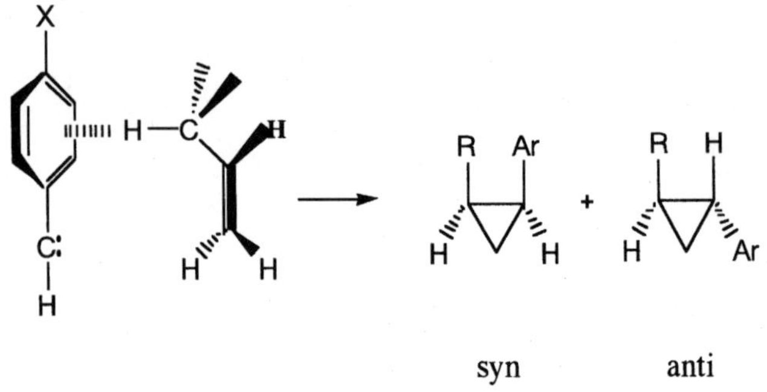

Figure 7.13. Addition of aryl carbenes to olefins.

TABLE 7.5. Stereochemical Results of the Coupling of 1,3-Bis(bromomethyl)-7-t-Butylpyrene with 1,3-Bis(mercaptomethyl)benzenes

		Yields (%)	
R^1	R^2	Anti	Syn
H	H	0	38
CH_3	t-C_4H_9	35	0
CH_2CH_3	t-C_4H_9	44	0
OCH_3	t-C_4H_9	0	25
F	t-C_4H_9	0	41

The anti isomer was obtained as a single product in the cases of 9-methyl and 9-ethyl compounds, whereas the syn conformer was the exclusive product for R^1 = H, OCH_3, and F.

A plausible explanation is that there is a compromise between the attractive and repulsive interactions. The interaction of CH_3 or CH_2CH_3 with the pyrene aromatic ring is favorable in terms of the CH/π interaction to give the anti isomer, while the interaction of OCH_3 and F versus the pyrene moiety is unfavorable in view of the dipolar interaction and leads to preferential formation of the syn product (Fig. 7.14).

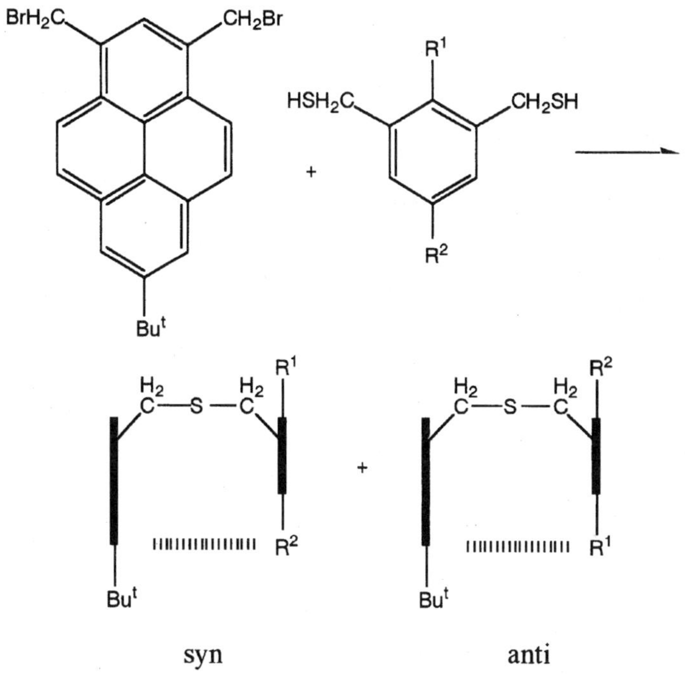

Figure 7.14. Coupling of 1,3-bis(bromomethyl)-7-*t*-butylpyrene with 1,3-bis(mercaptomethyl)benzenes.

REFERENCES

1. J. D. Morrison and H. S. Mosher, *Asymmetric Organic Reactions,* Prentice-Hall, NJ, 1971, Tables 2.1, 2.2.

2. E. J. Corey, K. B. Becker, and R. K. Varma, *J. Am. Chem. Soc.,* **94**, 8616 (1972).

3. K. Nishihata and M. Nishio, *Tetrahedron Lett.,* 1041 (1977).

4. M. Chérest, H. Felkin, and N. Prudent, *Tetrahedron Lett.,* 2199 (1968); M. Chérest and N. Prudent, *Tetrahedron,* **36**, 1599 (1980).

5. M. Hirota, K. Abe, H. Tashiro, and M. Nishio, *Chem. Lett.,* 777 (1982).

6. Y. Yokowo, T. Sakurai, M. Saburi, and S. Yoshikawa, *Nippon Kagaku Kaishi,* 1904 (1981).

7. S. Saito, Y. Morikawa, and T. Moriwake, *J. Org. Chem.,* **55**, 5424 (1990).

8. R. Breslow, R. Corcoran, J. A. Dales, S. Liu, and P. Kalicky, *J. Am. Chem. Soc.,* **96**, 1973 (1974); R. Breslow, R. J. Corcoran, and B. B. Snider, *ibid.,*

96, 6791 (1974); R. Breslow, B. B. Snider, and R. J. Corcoran, *ibid.*, **96**, 6792 (1974); B. B. Snider, R. J. Corcoran, and R. Breslow, *ibid.*, **97**, 6580 (1975).

9. R. Breslow, J. Rothbard, F. Herman, and M. Rodoriguez, *J. Am. Chem. Soc.*, **100**, 1213 (1978).

10. R. Breslow and L. M. Maresca, *Tetrahedron Lett.*, 887 (1978).

11. J. S. Birwistle, K. Lee, J. D. Morrison, W. A. Sanderson, and H. S. Mosher, *J. Org. Chem.*, **29**, 37 (1964); J. D. Morrison and H. S. Mosher, *Asymmetric Organic Reactions*, Prentice-Hall, NJ, 1971, Tables 5.6, 5.7.

12. J. Capillon and J. P. Guetté, *Tetrahedron*, **35**, 1817 (1979).

13. M. Chérest and N. Prudent, *Tetrahedron*, **36**, 1599 (1980).

14. Y. Kobuke, T. Fueno, and J. Furukawa, *J. Am. Chem. Soc.*, **92**, 6548 (1970).

15. G. L. Closs and R. A. Moss, *J. Am. Chem. Soc.*, **86**, 4042 (1964); R. A. Moss, *J. Org. Chem.*, **30**, 3261 (1965).

16. C. P. Casey, S. W. Polichnowski, A. J. Shusterman, and C. R. Jones, *J. Am. Chem. Soc.*, **101**, 7282 (1979).

17. T. Yamato, A. Miyazawa, and M. Tashiro, *J. Chem. Soc.*, *Perkin 1*, 3127 (1993).

CHAPTER 8

CRYSTAL STRUCTURES OF CLATHRATES

In Chapters 8 and 9 we examine CH/π interactions in supramolecular chemistry. Here, crystallographic data in lattice-tipe clathrates which possibly implicate the CH/π interaction are compiled.

Stoddart et al. reported that in the solid-state structure of a clathrate formed with tetraphenylborate anion (BPh_4^-) and paraquat dication **1**, the planar molecule **1** is sandwiched between two BPh_4^- anions.[1] An aromatic CH/π interaction has been suggested to be present between **1** and the orthogonally disposed phenyl rings on the anion BPh_4^-. The ^{1}H NMR spectral data showed that the close association between the ions is maintained in solution.

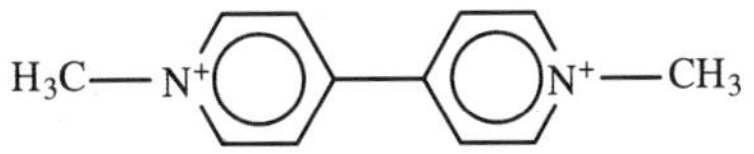

$$H_3C—N^+ \bigcirc \!\!\!\!\!\!\!\! \bigcirc N^+—CH_3$$

1

They also examined the solid state structure of diazabenzo-30-crown-10 disulfonamide **2**.[2] Inspecting the mode of packing of symmetry-related pairs of molecules revealed intermolecular interlocking between the N-tosyl groups, A and A' (Fig. 8.1), and the catechol residues, B and B'. Additionally, interaction was suggested to occur between rings A and B' (and rings A' and B); the centroid/

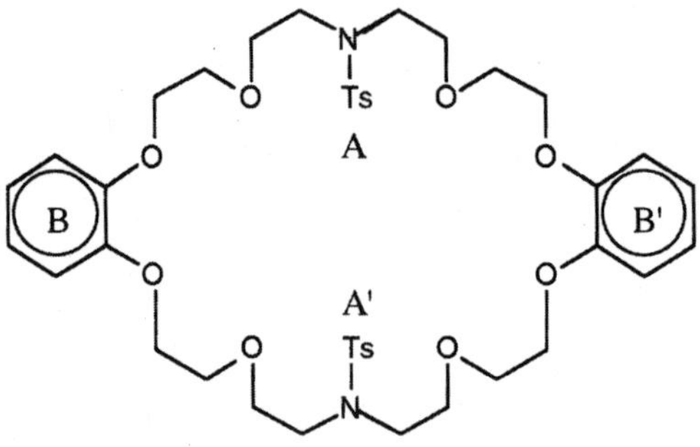

diazabenzo-30-crown-10 disulfonamide **2**

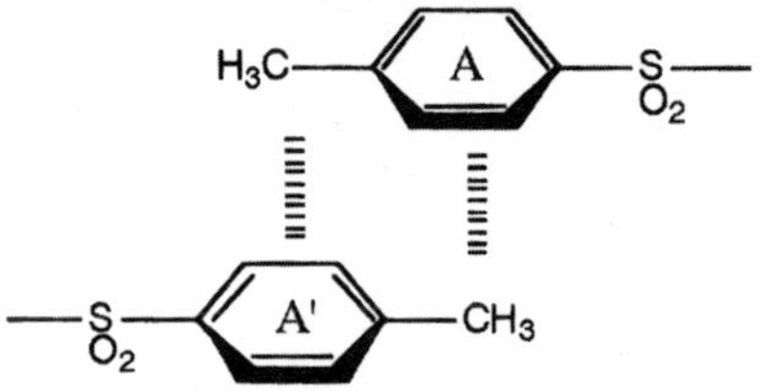

Figure 8.1. Interlocking between the *N*-tosyl groups in the crystal of diaza-benzo-30-crown-10 disulfonamide **2**.

centroid distance was 4.88 Å and the two ring mean planes inclined by 74° to each other.

Krieger and Diederich determined the crystal structures of 1′,1″-dimethyl-dispiro[1,6,20,25-tetraoxa[6.1.6.1]paracyclophane-13,4′:32,4″-bispiperidine] **3** in complex with benzene (**3**: benzene:water = 1:2:1) and *p*-xylene (**3**: *p*-xylene = 1:1).[3]

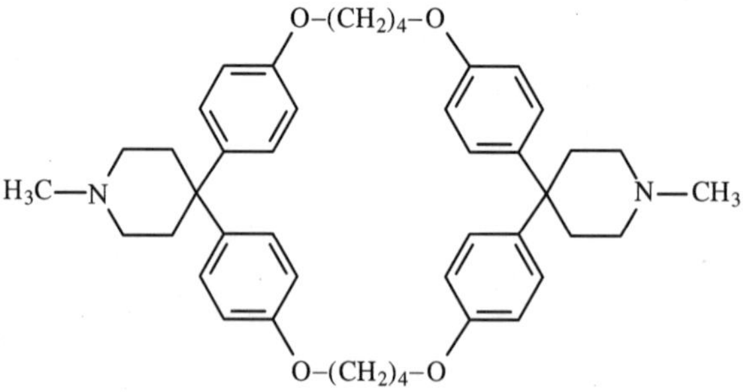

3

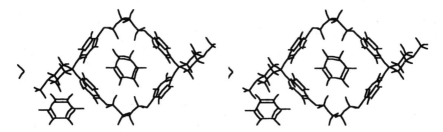

Figure 8.2. Crystal structure of a complex (**3**: benzene:water = 1:2:1), DENFOR (Refcode in the Cambridge Structural Database, CSD).

One benzene molecule is incorporated in the host molecule exactly in the center of the intramolecular cavity. The other benzene ring and water molecule are located outside the cavity between host molecules in the crystal lattice (Fig. 8.2).

Short C/C_{sp^2} contacts (3.45 and 3.62 Å) have been found between benzene and **3**. In complexes with toluene and *p*-xylene, the guest molecule is sandwiched between two adjacent macrocycles in the stack. The two methyl groups of a *p*-xylene molecule are inserted into the cavities of the two sandwiching hosts (Fig. 8.3). The shortest C/C_{sp^2} contact was 3.48 Å. It was noted that in these complexes the guests located in the intramolecular cavities almost exclusively interact with the aromatic rings of **3**.

Ogura et al. examined the inclusion of a number of alkyl aryl sulfoxides in crystals of a dipeptide, (*R*)-phenylglycyl-(*R*)-phenylglycine (Fig. 8.4).[4]

(*R*)-phenylglycyl-(*R*)-phenylglycine

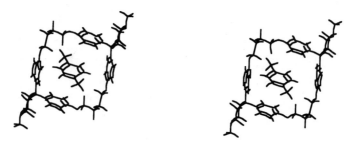

Figure 8.3. Crystal structure of a clathrate (**3**: p-xylene = 1:1), DENFUX.

(S)-Isopropyl phenyl sulfoxide formed a complex with the dipeptide in preference to the (R)-enantiomer. A number of CH/π contacts have been shown to exist between a methyl hydrogen (guest) and the aromatic nucleus of the host (2.7 Å) as well as between an aromatic CH (guest) and a sp^2 carbon of the host (3.0 and 3.2 Å). A search by CHPI (Chapter 11) unveiled five other contacts (3.0, 3.1, 3.1, 2.9, 3.1 Å), most of which are intermolecular. Ogura argued that this demonstrates the importance of the CH/π interaction in forming stable clathrates.[5]

Foces-Foces et al. studied the molecular structures of clathrates of benzene with 1,8-diaminonaphthalene **4** and 1-amino-8-triphenyl-phosphoranylideneaminonaphthalene **5**.[6]

$$NH_2 \quad NH_2$$

4

$$NH_2 \quad N{=}P(C_6H_5)_3$$

5

The intermolecular contacts involving CH atoms and the centroids of the fused benzene rings were found in both compounds (CH/C 2.62 and 2.96 Å, respectively, for **4** and **5**). In the latter compound, aromatic solvent molecules are also involved in this type of interaction. They are of a quasi T-type with distances between 4.6–4.9 Å and 4.9–5.1 Å, for **4** and **5**, respectively.

Hunter et al.[7] reported crystallographic evidence for the CH/π interaction in a clathrate formed by a bisphosphonium cyanoborate salt **6** with the solvent furan. The furan unit interacts with two hydrogen atoms from opposite faces of the plane of the

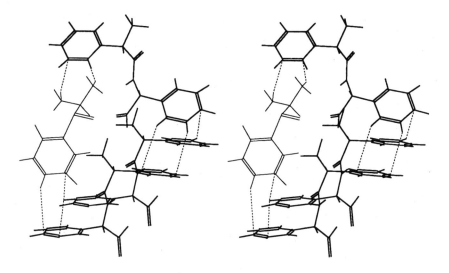

Figure 8.4. Stereo view of the crystal structure of (R)-phenylglycyl-(R)-phenylglycine (thick lines) complexed with (S)-isopropyl phenyl sulfoxide (thin lines).

heteroaromatic ring, which originate from a methyl group (2.67 Å above the aromatic ring plane) and from the ortho position of a phenyl group (2.79 Å) of different phosphonium ions.

$$Ph_3P^+Me \qquad\qquad Ph_3P^+Me$$

6

They studied[8] further the structures of clathrates formed by bis-phosphonium cyanoborate salt with p-xylene. The xylene guest molecules have been found to occupy constricted channels in the lattice and the shortest CH(methyl)/C(aromatic) distances are 2.8 and 2.9 Å. A search of the Cambridge Structural Database (CSD)[9] for interactions involving p-xylene as a guest revealed 11 other clathrates which display potential CH/π interactions. Mean values for 17 independent measurements of the perpendicular distance of the hydrogen atom from the plane of the aromatic ring is 2.8 Å. These distances compare with the corresponding values of 2.4 Å for OH/Ph and 2.5 Å for NH/Ph interactions.[10]

Mingos et al. studied the crystal structures of chloroform solvates of several gold ethyne complexes.[11] They found short atomic contacts between $CHCl_3$ and the π-atoms (2.45 and 2.55 Å) (Fig. 8.5). The CH/π interaction seems to be favored by the $CHCl_3$ proton and the donation of electrons by the Au atoms into the triple bond. A search of CSD for interactions involving chloroform revealed nine other organometallic and organic compounds which display potential CH/π(C≡C) interactions with similar dimensions; the CH bonds are directed nearly orthogonally toward the π-surfaces in every case.

Steiner found a short CH/π contact (2.54 Å) in crystals of DL-prop-2-ynylglycine.[12] The prop-2-ynyl residue points almost linearly at the terminal C atom of a symmetrically related prop-2-ynyl residue, whereby an infinite chain of C≡C–H ⋯ C≡C–H contacts is formed (Fig. 8.6). Occurrence of similar contacts in other crystal structures has been checked by a search through CSD. A number of short contacts has been identified; the contacts are not isolated, but form zigzag patterns as in Fig. 8.6.

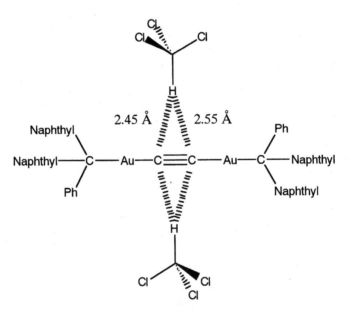

Figure 8.5. Crystal structure of the chloroform solvate of a gold ethyne complex.

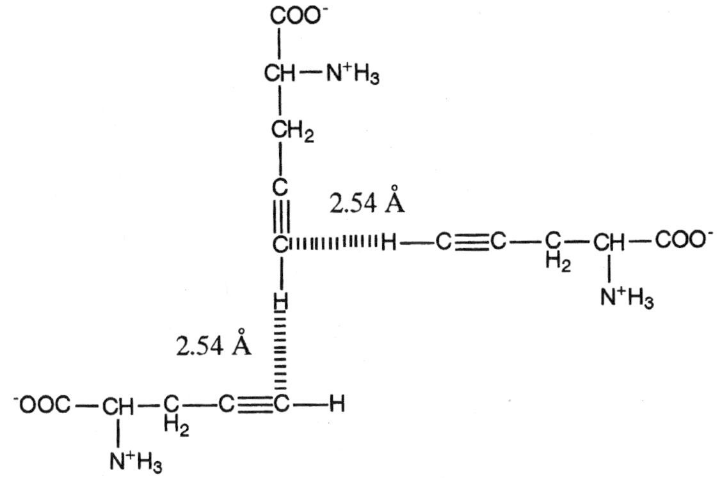

COO⁻

CH—N⁺H₃

CH₂

C

2.54 Å

C""""""""H—C≡C—C—CH—COO⁻
                        H₂
                              N⁺H₃

H

2.54 Å

⁻OOC—CH—C—C≡C—H
         H₂
N⁺H₃

Figure 8.6. DL-Prop-2-ynylglycine.

An example of a cooperative OH/π and CH/π interaction has been given by the same authors[13] in the solid state dimer structure of 7-ethynyl-6,8-diphenyl-7H-benzocyclohepten-7-ol. Kobayashi et al. also reported CH/π interactions in clathrates of 1-(9-anthryloxy)-anthraquinone with benzene, anthracene, and hydroquinone.[14]

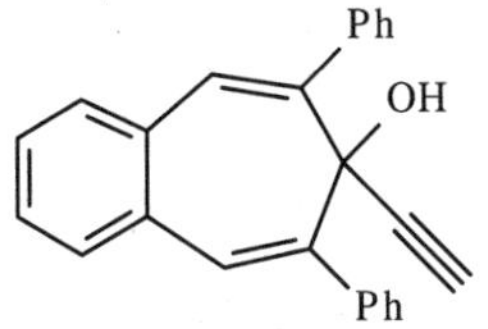

7-ethynyl-6,8-diphenyl-7 *H*-benzocyclohepten-7-ol

Bürgi et al. reported the crystal structure of tris(bicyclo-[2,1,1]-hexeno)benzene.[15] The center of a molecule in one layer has been found to be situated over a hydrogen atom in the wing (aromatic moiety) of a molecule in a neighboring layer. The juxtaposition of CH against the aromatic ring suggested the possibility of a CH/π interaction.

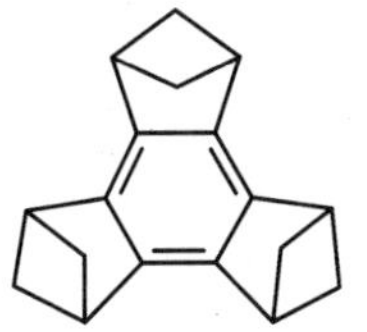

tris(bicyclo-[2,1,1]hexeno)benzene

Siegel et al. studied the crystal structure of trisbicyclo-[2,1,1]-heptabenzene.[16] The distances between the mean plane of the ring and the methylene hydrogens are 3.14 and 4.08 Å. Based on these results, they commented on the possibility of a weak CH/π interaction.

trisbicyclo-[2,1,1]heptabenzene

Methanol, ethanol, and 1-propanol were reported to form clathrates with 2-[o-(triphenylphosphoranylidenamino)benzyliden]amino-1-H-2,3-dihydroindazol-3-one.[17] Formation of a methanol clathrate with $\alpha,\alpha,\alpha',\alpha'$-tetraphenyl-1,3-dioxolane-4,5-dimethanol was reported.[18] Short interatomic distances have been recorded between a carbon of the guests and the aromatic carbon of the host. Short contacts between a methyl group with an aromatic moiety have also been shown in the crystal structures of various clathrates.[19]

REFERENCES

1. G. J. Moody, R. K. Owusu, A. M. Z. Slawin, N. Spencer, J. F. Stoddart, J. D. R. Thomas, and D. J. Williams, *Angew. Chem. Int. Ed.*, **26**, 890 (1987).

2. P. L. Anelli, A. M. Z. Slawin, J. F. Stoddart, and D. J. Williams, *Tetrahedron Lett.*, 1575 (1988); P. L. Anelli, J. F. Stoddart, A. M. Z. Slawin, and D. J. Williams, *Acta Cryst. C*, **46**, 1468 (1990). See also D. B. Amabilino, P. R. Ashton, C. L. Brown, E. Cordova, L. A. Godinez, T. T. Goodnow, A. E. Kaifer, S. P. Newton, M. Pietraszkiewicz, D. Philp, F. M. Raymo, A. S. Reder, M. T. Rutland, A. M. Z. Slawin, N. Spencer, J. F. Stoddart, and D. J. Williams, *J. Am. Chem. Soc.*, **117**, 1271 (1995).

3. C. Krieger and F. Diederich, *Chem. Ber.*, **118**, 3620 (1985); Review: F. Diederich, *Angew. Chem. Int. Ed.*, **27**, 362 (1988).

4. K. Ogura, T. Uchida, M. Noguchi, M. Minoguchi, A. Murata, M. Fujita, and K. Ogata, *Tetrahedron Lett.*, 3331 (1990); M. Akazome, M. Noguchi, O. Tanaka, A. Sumikawa, T. Uchida, and K. Ogura, *Tetrahedron*, **53**, 8315 (1997).

5. Review: K. Ogura, *Yukagaku*, **43**, 779 (1994).

6. A. L. Llamas-Saiz, C. Foces-Foces, P. Molina, M. Alajarin, A. Vidal, R. M. Claramunt, and J. Elgueno, *J. Chem. Soc., Perkin 2*, 1025 (1991).

7. R. Hunter, R. H. Haueisen, and A. Irving, *Angew. Chem. Int. Ed.*, **33**, 566 (1994).

8. A. Irving, R. Hunter, and R. H. Haueisen, *J. Chem. Crystallogr.*, **24**, 267 (1994).

9. F. H. Allen, J. E. Davies, J. J. Galloy, O. Johnson, O. Kennard, C. F. Macrae, E. M. Mitchell, G. F. Mitchell, J. M. Smith, and D. G. Watson, *J. Chem. Inf. Comput. Sci.*, **31**, 187 (1990).

10. M. A. Viswamitra, R. Radhakrishnan, J. Bandekar, and G. R. Desiraju, *J. Am. Chem. Soc.*, **115**, 4868 (1993).

11. T. E. Müller, D. M. P. Mingos, and D. J. Williams, *J. Chem. Soc., Chem. Commun.*, 1787 (1994).

12. T. Steiner, *J. Chem. Soc., Chem. Commun.*, 95 (1995).

13. T. Steiner, T. Tamm, B. Lutz, and J. van der Maas, *Chem. Commun.*, 1127 (1996); T. Steiner, E. B. Starikov, A. M. Amado, and J. J. C. Teixeira-Dias, *J. Chem. Soc., Perkin 2*, 1321 (1995); T. Steiner, E. B. Starikov, and M. Tamm, *ibid.*, 67 (1996).

14. K. Ochiai, Y. Mazaki, and K. Kobayashi, *Tetrahedron Lett.*, 5947 (1995); K. Ochiai, Y. Mazaki, S. Nishikiori, K. Kobayashi, and S. Hayashi, *J. Chem. Soc., Perkin 2*, 1139 (1996).

15. H.-B. Bürgi, K. K. Baldridge, K. Hardcastle, N. L. Frank, P. Gantzel, J. S. Siegel, and J. Ziller, *Angew. Chem. Int. Ed.*, **34**, 1454 (1995).

16. N. L. Frank, K. K. Baldridge, P. Gantzel, and J. S. Siegel, *Tetrahedron Lett.*, **25**, 4389 (1995).

17. P. Molina, A. Arques, P. Obon, A. L. Llamas-Saiz, C. Foces-Foces, R. M. Claramunt, C. Lopez, and J. Elguero, *J. Phys. Org. Chem.*, **5**, 507 (1992).

18. E. Weber, N. Dörpinghaus, C. Wimmer, Z. Stein, H. Krupitsky, and I. Goldberg, *J. Org. Chem.*, **57**, 6825 (1992); I. Csöregh, O. Gallardo, E. Weber, M. Hecker, and A. Wierig, *J. Chem. Soc., Perkin 2*, 1939 (1992).

19. A. Uchida, Y. Ohashi, Y. Sasada, M. Moriya, and T. Endo, *Acta Cryst. C*, **40**, 120 (1984); A. Petti, T. J. Shepodd, R. E. Barrans, and D. A. Dougherty, *J. Am. Chem. Soc.*, **110**, 6825 (1988); X. Wang, S. D. Erickson, T. Iimori, and W. C. Still, *ibid.*, **114**, 4128 (1992); M. E. Tanner, C. B. Knobler, and D. J. Cram, *J. Org. Chem.*, **57**, 40 (1992); M. Roos and J. L. M. Dillen, *Acta Cryst. C*, **48**, 1882 (1992); B. Hinzen, P. Seiler, and F. Diederich, *Helv. Chim. Acta* **79**, 942 (1996); F.-G. Klärner, J. Benkhoff, R. Boese, U. Burkert, M. Kamieth, and U. Naatz, *Angew. Chem. Int. Ed.*, **35**, 1130 (1996).

CHAPTER 9

INCLUSION COMPOUNDS

In this chapter we examine crystallographic, spectroscopic, and thermodynamic data of host–guest complexes. Inclusion compounds of cyclodextrins and synthetic macrocycles such as calixarenes, cavitands, and cyclophanes are mentioned.

9.1. CYCLODEXTRIN COMPLEXES

The thermodynamics of complex formation was studied for hexakis(2,6-di-O-methyl)-α-cyclodextrin (CD) with p- and m-substituted benzoic acids, phenols, and anilines (X-C$_6$H$_4$-Y; X = CO$_2$H, OH, or NH$_2$; Y = H, OH, NH$_2$, or NO$_2$), by measurement of induced circular dichroism spectra (Table 9.1).[1]

Negative values were obtained for ΔH and ΔS. The results have been interpreted to indicate that the complex is formed by the tight binding of the guests with the host CD. In other words, the driving force of complex formation is enthalpic in origin and is by no means regulated by the so-called hydrophobic interaction. Short CH/C$_{sp^2}$ distances are in fact found among the X-ray crystallographic data of cyclodextrin (CD) complexes (Table 9.2).[2]

TABLE 9.1. Thermodynamic Parameters of α-Dimethyl CD with Various Benzene Derivatives X-C$_6$H$_4$-Y (H$_2$O at 298 K)

X	Y	ΔG (kcal mol^{-1})	ΔH (kcal mol^{-1})	ΔS (K^{-1} cal mol^{-1})
CO$_2$H	H	− 3.89	− 14.3	− 34.9
CO$_2$H	m-OH	− 3.65	− 13.6	− 33.5
CO$_2$H	p-OH	− 4.09	− 11.6	− 25.1
CO$_2$H	m-NH$_2$	− 4.39	− 13.6	− 30.8
OH	m-NO$_2$	− 3.48	− 7.8	− 14.5
OH	p-NO$_2$	− 3.59	− 6.7	− 10.3
NH$_2$	m-NO$_2$	− 3.66	− 4.4	− 2.5
NH$_2$	p-NO$_2$	− 4.40	− 9.7	− 17.8

TABLE 9.2. X-ray Crystallographic Data of Cyclodextrin Complexes

α-CD/m-Nitroaniline, 1-phenylethanol[3]
β-CD/Nicotineamide[4]
β-CD/Fenoprofen[5]
β-CD/m-Iodophenol, biphenylacetic acid[6]
β-CD/2-Naphthoic acid[7]
Trimethyl-β-CD/p-iodophenol[8]
Trimethyl-β-CD/p-iodoaniline, benzaldehyde, p-nitrophenol, flurbiprofen[9]

(R)-(−)-fenoprofen   flurbiprofen

9.2. CALIXARENE COMPLEXES

The complexation of a series of alcohols with a calixarene compound **1** [X = H, R = (CH$_2$)$_{10}$CH$_3$] was studied.[10] The complexation-induced chemical shifts (CIS) of protons in 2-pentanol increased in the order 1-CH$_2$ and 3-CH$_2$ (δ 1.4–1.5) < 4-CH$_2$ (δ 1.6) < 5-CH$_3$

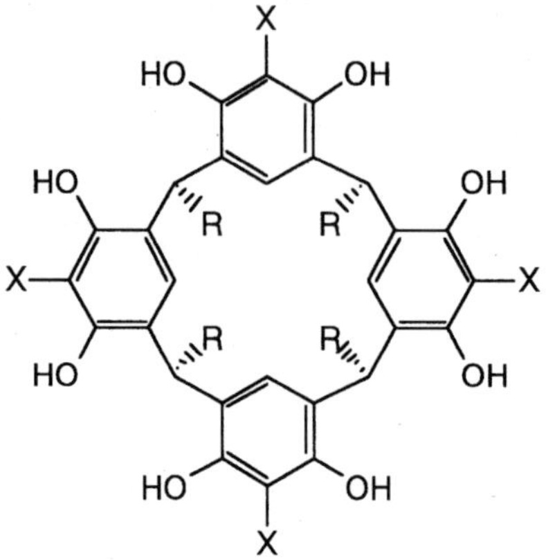

1

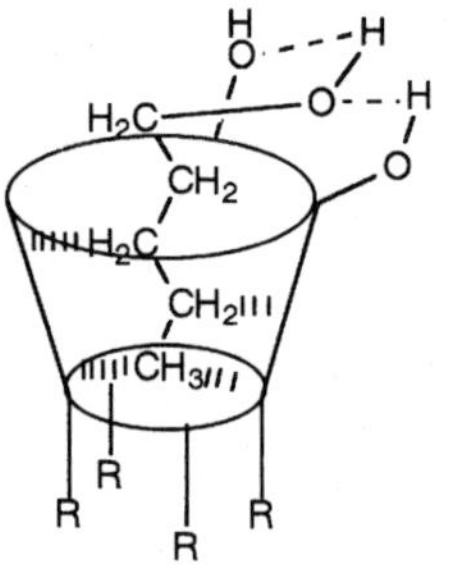

Figure 9.1. Complexation of 2-pentanol with a calixarene derivative.

(δ 1.8–1.9). The otherwise flexibly moving 5-methyl group in 2-pentanol was suggested to swing over into the aromatic cavity of **1** (Fig. 9.1).

Alcohols investigated included 1-propanol (**a**), 1-butanol (**b**), *t*-butanol (**c**), neopentyl alcohol (**d**), and 3,3-dimethyl-1-butanol (**e**). The binding ability of the alcohols with **1** increased with increasing chain length of the alkyl group and the variation in K_{rel} (relative strength of the complexation). This parallels the results with CISs for the terminal methyl groups (Table 9.3). The trend above is understood in terms of the CH/π interaction, since the extent of the interaction becomes greater as the number of CH groups involved in the complexation increases.

Crystallographic,[11] spectroscopic, and thermodynamic data which possibly implicate the CH/π interaction are found in calixarene

TABLE 9.3. Relative Strength of the Complexation of 1 and CIS^a with Alcohols (CDCl$_3$ at 298 K)

Guest	a	b	c	d	e
K_{rel}	0.7	1.3	0.7	1	4.3
$\Delta\delta CH_3$ (ppm)	1.15	1.47	1.47	1.82	1.95

^{a} A positive value corresponds to a high-field shift.

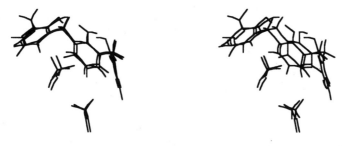

Figure 9.2. Crystal structure of a complex, **2**: acetone (BIMXIE: Refcode in the CSD).[31]

complexes (**1–3**) listed in Table 9.4. Figure 9.2 illustrates an X-ray result (**2**, R^1 = OH, R^2 = H)/acetone.

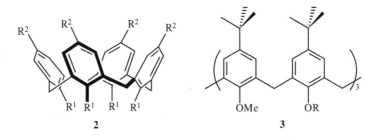

9.3. CAVIPLEXES

In caviplexes **4–8** are found in crystallographic and spectroscopic data which may implicate the CH/π interaction (Table 9.5). Figure 9.3 illustrates an X-ray result (**5**, R = C$_6$H$_{13}$/fluorobenzene).

TABLE 9.4. X-ray, Spectroscopic, and Thermodynamic Data of Calixarene Complexes[a]

---

2 (R^1 = OH, R^2 = Bu^t)/toluene (X-ray,[12] NMR[13])

2 (R^1 = OH, R^2 = 1,1,3,3-tetramethylbutyl)/toluene (X-ray)[14]

2 (R^1 = OH, R^2 = H/acetone (X-ray)[15] (Fig. 9.2)

2 (R^1 = OH, R^2 = Bu^t)/anisole (X-ray)[16]

2 (R^1 = CO$_2$C$_2$H$_5$, R^2 = Bu^t)/acetonitrile (X-ray)[17]

2 (R^1 = OH, R^2 = Bu^t)/pyridine (X-ray)[18]

2 (R^1 = CO$_2$C$_2$H$_5$, R^2 = Bu^t)/ethanol (X-ray)[19]

2 (R^1 = OH, R^2 = Pr^i)/p-xylene (X-ray, DSC)[20]

Aluminium-fused bis-p-t-butylcalix[4]arene/methylene chloride (X-ray)[21]

1 [X = H; R = CH$_2$)$_{10}$CH$_3$]/carbohydrates (NMR, CD)[22]

1 (X = H, OH, Me; R = CH$_2$CH$_2$SO$_3$Na)/alcohols, carbohydrates, nucleosides (NMR)[23]

1 (X = H, OH, Me; R = CH$_2$CH$_2$SO$_3$Na)/acetyl choline, etc. (thermodynamic)[24]

Calix[6]arene/RNH$_3^+$ (R = n-C$_8$H$_{17}$, CH$_2$CH$_2$C$_6$H$_5$, CH$_2$C$_6$H$_5$, adamantyl) (thermodynamic)[25]

1 (X = H, OH; R = CH$_2$CH$_2$SO$_3$Na)/amino acids (NMR, thermodynamic)[26]

1, **1**$^{2-}$, **1**$^{4-}$ (deprotonated forms) (X = H, OH, Me; R = CH$_2$CH$_2$SO$_3$Na)/ fucose, cyclohexanol, furan (NMR, thermodynamic)[27]

2,4,6-Trisubstituted calix[6]arenes **3** (R = Me, Et, n-Pr) (NMR)[28]

1 (X = H, OH, Me; R = CH$_2$CH$_2$SO$_3$Na)/various hydrophilic compounds (thermodynamic)[29]

1 (X = H; R = (CH$_2$)$_{10}$CH$_3$)/alkylbenzenes, alkyl benzoates (CD, thermodynamic)[30]

---

[a] X-ray, X-ray crystallography; NMR, ^{1}H-NMR; DSC, differential scanning calorimetry; CD, circular dichroism; thermodynamic, consideration on formation constants.

9.4. CYCLOPHANE COMPLEXES

Crystallographic, spectroscopic, and thermodynamic data which may implicate the CH/π interaction are found in the cyclophane and cryptophane complexes **9–18** listed in Table 9.6.[39] Figure 9.4 is the crystal structure of 1,6,20,25-tetraaza[6.1.6.1]paracyclophane **9** complexed with durene (1,2,4,5-tetramethylbenzene). Favorable methyl/aromatic-ring interaction has also been noted for a methyl ammonium complex with a speleand.[40]

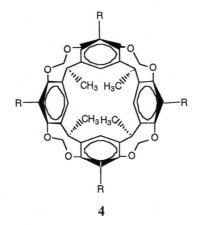

4

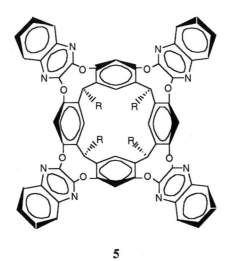

5

6

7

8

TABLE 9.5. X-ray and Spectroscopic Data of Caviplexes[a]

4 (R = CH$_3$, I)/acetonitrile, benzene, cyclohexane, toluene (X-ray)[32]
5 (R = C$_7$H$_{15}$)/aromatic compounds (X-ray, NMR)[33]
5 (R = C$_6$H$_{13}$)/benzene, butanol, etc. (MS)[34]
5 (R = C$_6$H$_{13}$)/acetone, fluorobenzene (X-ray, NMR)[35] (Fig. 9.3)
6 (R = C$_3$H$_7$)/methylenechloride (X-ray)[36]
7 (R = C$_2$H$_4$OC$_2$H$_5$)/ethyl acetate, acetonitrile, etc. (MS)[37]
8 (R = C$_2$H$_4$OC$_2$H$_5$)/prednisolone-21-acetate (NMR)[38]

[a] X-ray, X-ray crystallography; NMR, ^{1}H-NMR; MS, mass spectrometry.

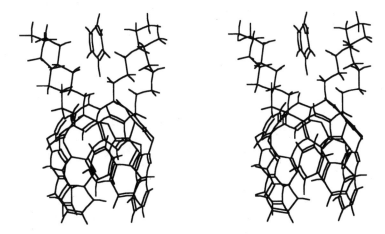

Figure 9.3. Crystal structure of a complex, **5**/fluorobenzene (R = C$_6$H$_{13}$, YAGVIL: Refcode in the CSD).

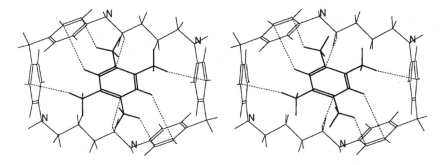

Figure 9.4. Crystal structure of 1,6,20,25-tetraaza[6.1.6.1]paracyclophane complexed with 1,2,4,5-tetramethylbenzene (stereo view). Dotted lines indicate short CH/C$_{sp^2}$ contacts.

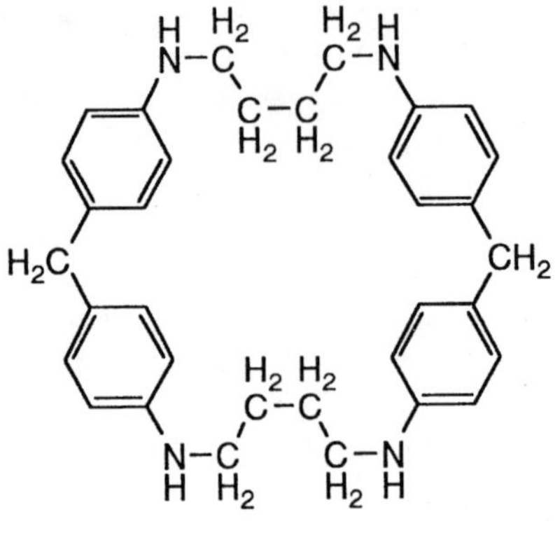

9

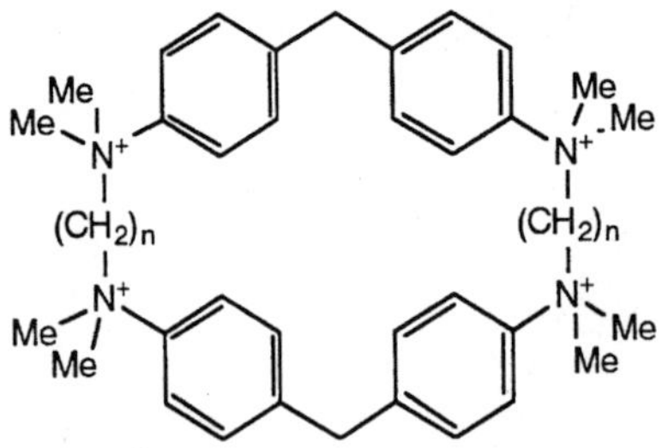

10 (n = 4, 5, 6)

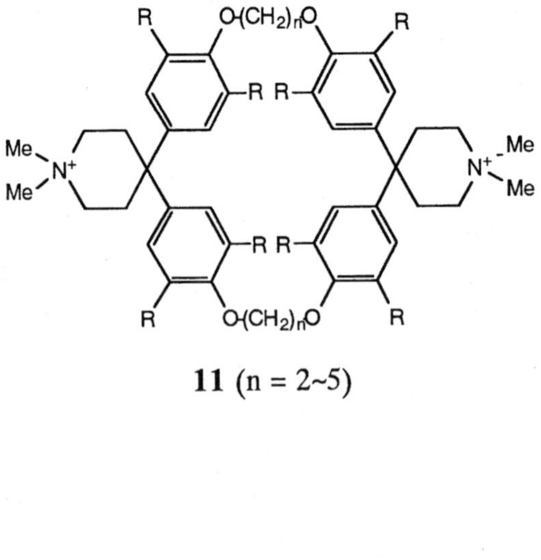

11 (n = 2~5)

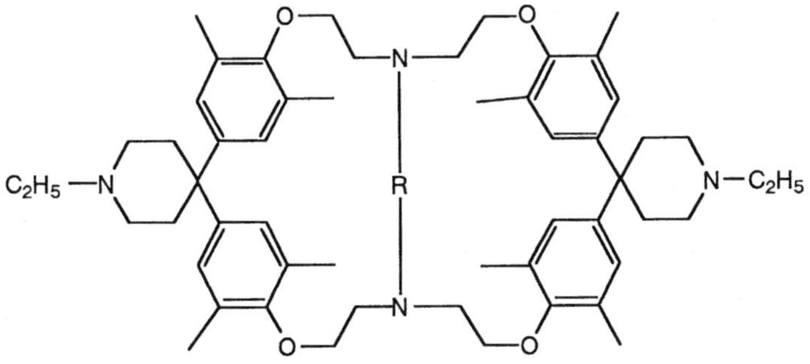

12

R =

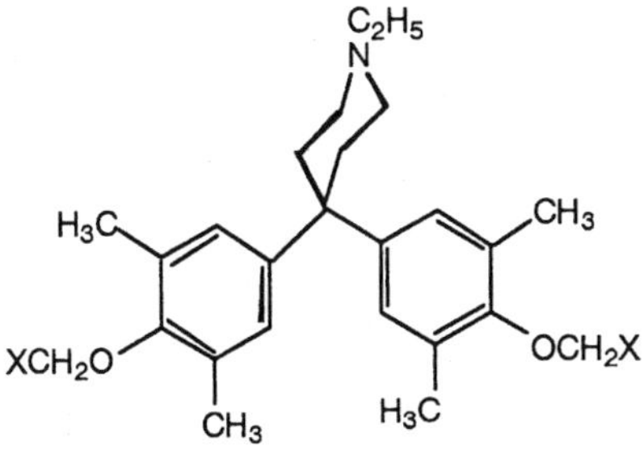

X = CH$_2$, CO

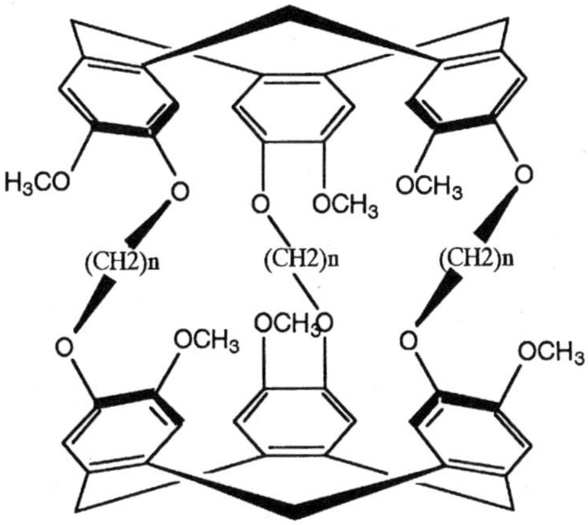

13 (n = 2, 3)

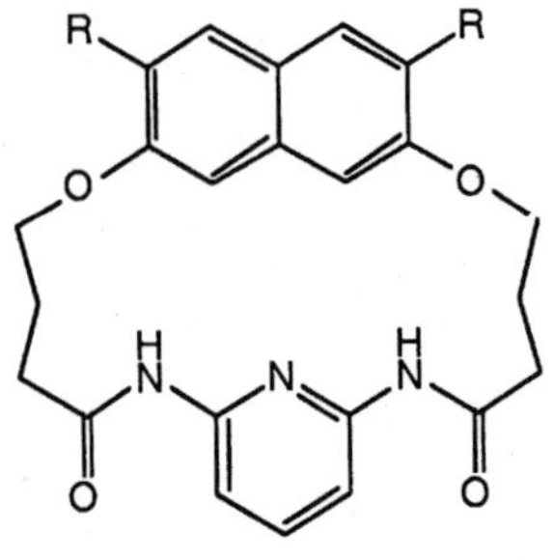

14 [R = O(CH2)3CH3]   1-butylthymine

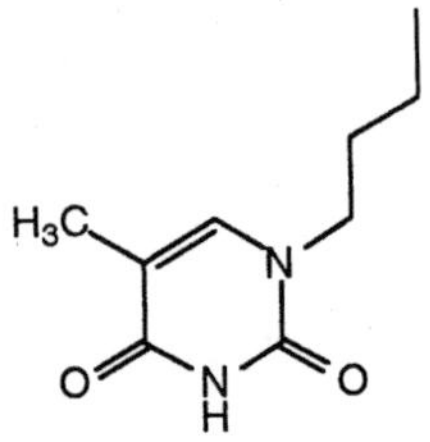

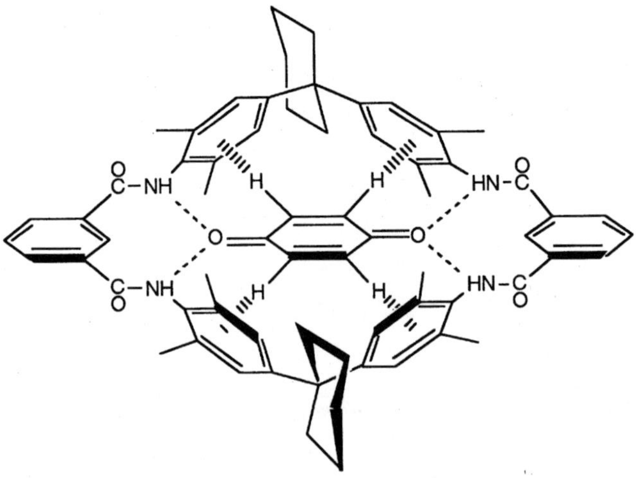

15

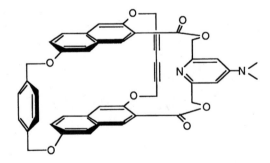

16

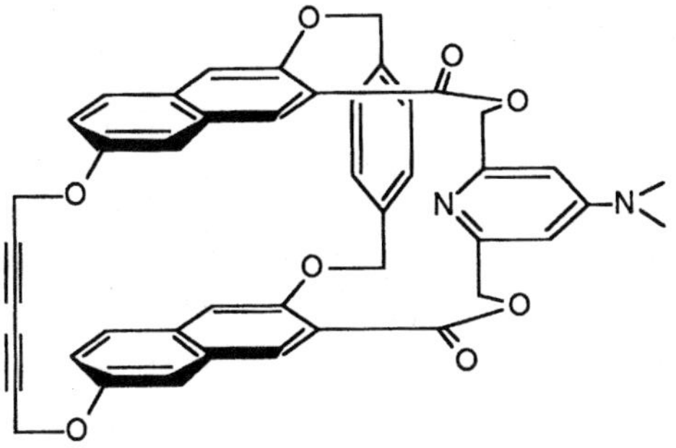

17

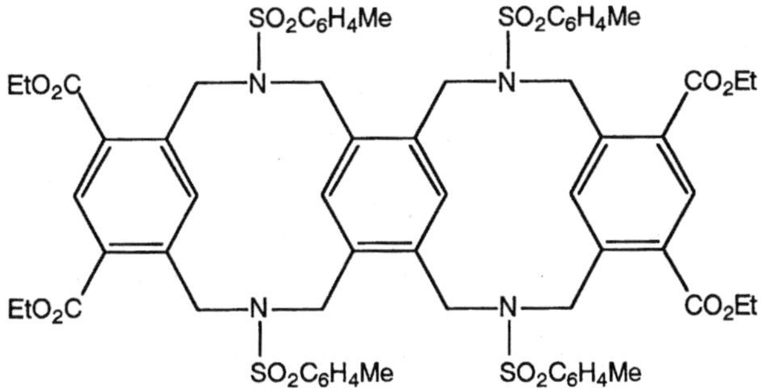

18

**TABLE 9.6. X-ray, Spectroscopic, and Thermodynamic Data of
Cyclophane Complexes**

9/durene (X-ray)[41]
9/naphthalene derivatives (NMR,[42] X-ray[43])
10/amino acids, nucleic acid bases, steroids (NMR, thermodynamic)[44]
11/benzene derivatives (NMR, thermodynamic)[45]
12/perylene, pyrene, fluoranthene (NMR, UV, thermodynamic)[46]
13/halogenated methanes (NMR, thermodynamic)[47]
14/1-butylthymine (X-ray)[48]
15/p-benzoquinone (NMR)[49]
16 and 17/p-nitrophenol (X-ray, NMR, thermodynamic)[50]
18/toluene (X-ray)[51]

9.5. PSEUDOROTAXANES

On addition to an acetonitrile solution of cyclobis(paraquat-*p*-pheny-lene)cyclophane **19**, compound **20**, bearing three 1,5-dioxynaphtha-lene residues, was included immediately to form a stable 1:1 complex, the formation constant K being 11100 mol^{-1} ($\Delta G°$ ca. 5.6 kcal mol^{-1}).[52] Large ^{1}H NMR chemical shift changes ($\Delta\delta - 5.41$ ppm) were observed for the aromatic protons H^4, H^8. An X-ray determination of pseudorotaxane demonstrated that there are pairs of CH/C$_{sp^2}$ contacts between H^4, H^8 of **20** and the *p*-xylyl bridging unit of **19** (Fig. 9.5),

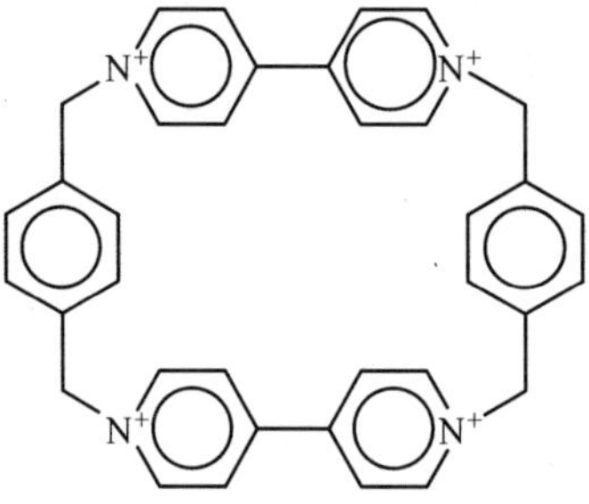

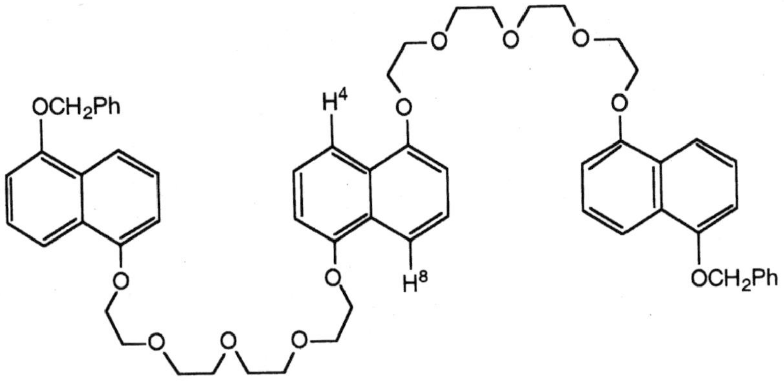

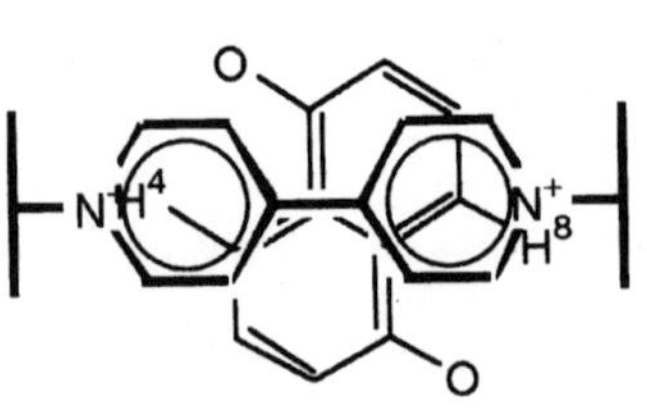

Figure 9.5. Schematic illustration of the structure of a pseudorotaxane composed of **20** and cyclobis(paraquat-*p*-phenylene)cyclophane **19**.

TABLE 9.7. X-ray Crystallographic Data of Pseudorotaxanes

19/1,3-Bis(5-hydroxy-1-naphthyloxy)propane (X-ray, NMR, thermodynamic)[53]
19/21 (X-ray)[54]
Bis-*p*-phenylene-37-crown-11/bipyridinium dication **22** (X-ray, NMR, thermodynamic)[55]
Bis-*p*-phenylene-32-crown-10/**23** (1:2) (X-ray)[56]

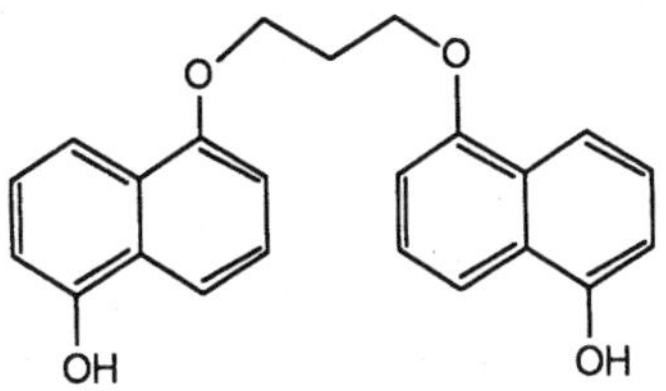

1,3-bis(5-hydroxy-1-naphthyloxy) propane

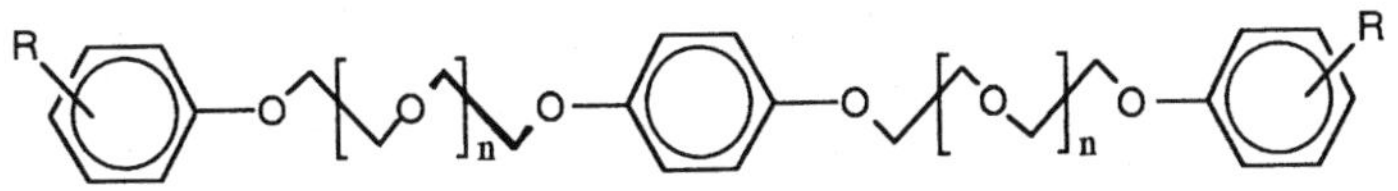

21 (n = 3, R = 4-PhCH2O; n = 2, R = 3,5-(MeO)2]

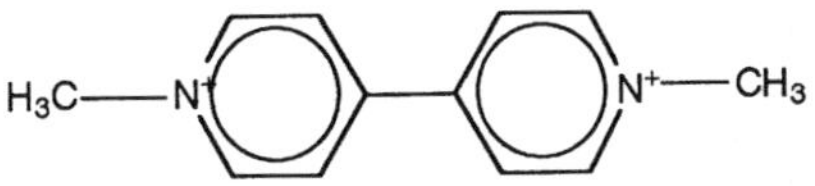

22

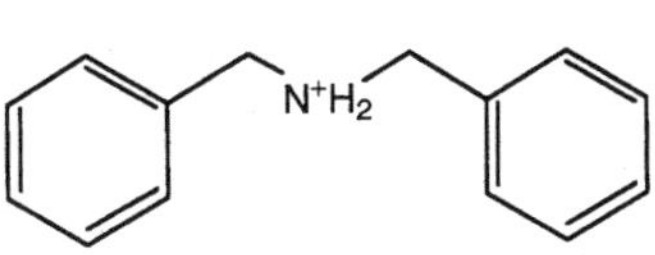

23

CH/ring-centroid distances, 2.54 Å). In the crystal lattice, the pseudo-rotaxanes are arranged to form infinite two-dimensional sheets, sustained by a combination of CH/π interactions and π/π stacking.

Crystallographic, spectroscopic, and thermodynamic data of pseudo-rotaxanes which possibly implicate the CH/π interaction are summarized in Table 9.7.

9.6. CATENANES

A series of [2]catenanes[57] composed of crowns and cyclobis(para-quat-*p*-phenylene)cyclophane **19** were prepared (Fig. 9.6).[58] The procedure involves insertion of the dicationic moiety into the macrocycle **24–26** along with folding of the termini of the paraquat reagent. The efficiency of the reaction is remarkable (Table 9.8).[59] This is certainly a consequence of arranging the reacting species in such a way as to afford an efficient catenation. The importance of π/π stacking and CH$/\pi$ and CH/O interactions was invoked to explain such a significant result. Support for this has been found among the X-ray, NMR, and thermodynamic data of the products including thiophene-containing [2]catenanes,[60] [*n*]catenanes (n: 3–5), catenated cyclodextrins,[61] bis[2]catenanes,[62] and [2]rotaxanes.[63]

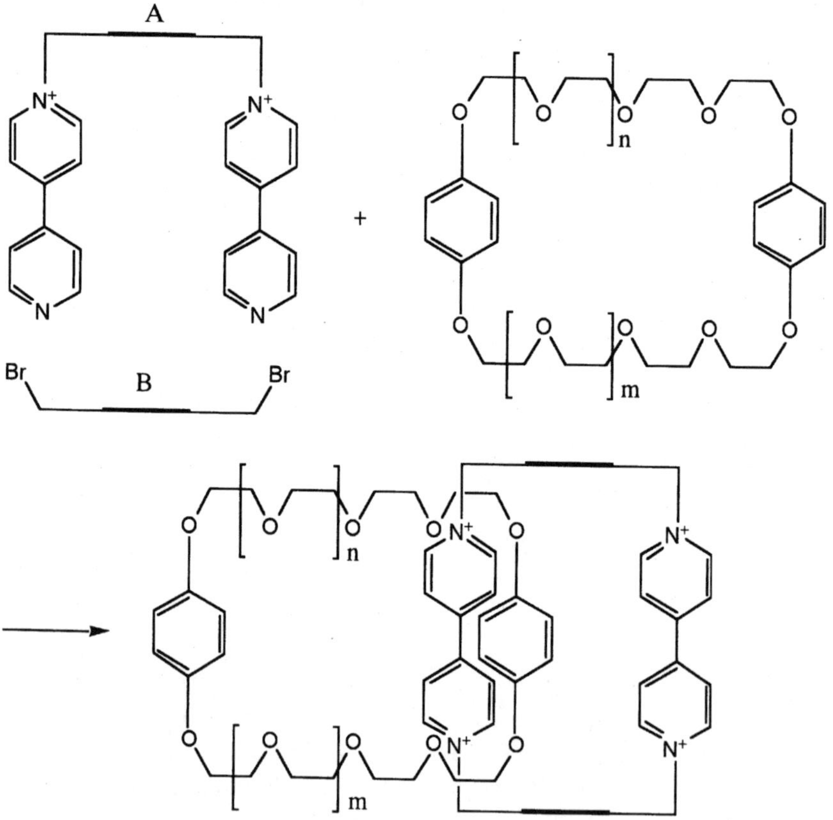

Figure 9.6. Syntheses of catenanes.

TABLE 9.8. Yields of [2]catenanes composed of crowns (24–26) and cyclobis(paraquat-p-phenylene)cyclophane 19

Components	A	B	Crown	Solvent	Yield (%)
	para	para	**24**	MeCN	70
	meta	para	**24**	DMF	40
	para	para	**25**	DMF	15
	meta	para	**25**	DMF	27
	para	para	**26**	DMF	17
	meta	para	**26**	DMF	12

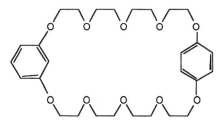

para =   meta =

bis-*p*-phenylene-34-crown-10 **24**

p-phenylene-*m*-phenylene-33-crown-10 **25**

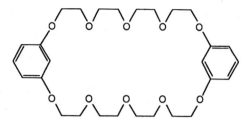

bis-*m*-phenylene-32-crown-10 **26**

REFERENCES

1. K. Harata, *J. Incl. Phenom.*, **13**, 177 (1992).
2. Review: K. Harata, *Inclusion Compounds*, Vol. 5, J. L. Atwood, J. E. D. Davies, and D. D. McNicol, Eds. (1991).
3. K. Harata, *Bull. Chem. Soc. Jpn.*, **53**, 2782 (1980); *ibid.*, **55**, 1367 (1982).
4. K. Harata, K. Kawano, K. Fukunaga, and Y. Ohtani, *Chem. Pharm. Bull.*, **31**, 1428 (1983).
5. J. A. Hamilton and L. Chen, *J. Am. Chem. Soc.*, **110**, 4379 (1988).
6. K. Harata, *J. Chem. Soc., Chem. Commun.*, 546 (1993).
7. K. Harata, F. Hirayama, H. Arima, K. Uekama, and T. Miyaji, *J. Chem. Soc., Perkin 2*, 1159 (1992).
8. K. Harata, K. Uekama, M. Otagiri, and F. Hirayama, *Bull. Chem. Soc. Jpn.*, **56**, 1732 (1983).
9. K. Harata, K. Uekama, M. Otagiri, and F. Hirayama, *J. Incl. Phenom.*, **1**, 279 (1984).
10. K. Kobayashi, Y. Asakawa, Y. Kikuchi, H. Toi, and Y. Aoyama, *J. Am. Chem. Soc.*, **115**, 2648 (1993).
11. Review: G. D. Andreetti, F. Ugozzoli, R. Ungaro, and A. Pochini, *Inclusion Compounds*, Vol. 4, J. L. Atwood, J. E. D. Davies, and D. D. McNicol, Eds., Oxford, 1991, p. 64; G. D. Andreetti and F. Ugozzoli, *Calixarenes: A Versatile Class of Macrocyclic Compounds*, Vol. 3, J. Vicens and V. Böhmer, Eds., Kluwer, Dordrecht, 1991, p. 87.
12. G. D. Andreetti, A. Pochini, and R. Ungaro, *J. Chem. Soc., Chem. Commun.*, 1005 (1979); S. G. Bott, A. W. Coleman, and J. L. Atwood, *J. Am. Chem. Soc.*, **108**, 1709 (1986).

13. T. Komoto, I. Ando, Y. Nakamoto, and S. Ishida, *J. Chem. Soc., Chem. Commun.*, 135 (1988).

14. G. D. Andretti, A. Pochini, and R. Ungaro, *J. Chem. Soc., Perkin 2*, 1773 (1983).

15. R. Ungaro, A. Pochini, G. D. Andreetti, and V. Sangermano, *J. Chem. Soc., Perkin 2*, 1979 (1984).

16. R. Ungaro, A. Pochini, G. D. Andreetti, and P. Domiano, *J. Chem. Soc., Perkin 2*, 197 (1985).

17. M. A. McKerby, E. M. Seward, G. Ferguson, and B. L. Rhul, *J. Org. Chem.*, **51**, 3581 (1986).

18. G. D. Andreetti, O. Ori, F. Ugozzoli, A. Alfieri, A. Pochini, and R. Ungaro, *J. Incl. Phenom.*, **6**, 523 (1988).

19. G. Ferguson, J. F. Gallagher, and S. Pappalardo, *J. Incl. Phenom.*, **14**, 349 (1992).

20. M. Perrin, F. Gharnati, D. Oehler, R. Perrin, and S. Lecocq, *J. Incl. Phenom.*, **14**, 257 (1992).

21. J. L. Atwood, S. G. Bott, C. Jones, and C. L. Ratson, *J. Chem. Soc., Chem. Commun.*, 1349 (1992).

22. Y. Kikuchi, Y. Tanaka, S. Sutarto, K. Kobayashi, H. Toi, and Y. Aoyama, *J. Am. Chem. Soc.*, **114**, 10302 (1992).

23. K. Kobayashi, Y. Asakawa, Y. Kato, and Y. Aoyama, *J. Am. Chem. Soc.*, **114**, 10307 (1992).

24. K. Kobayashi, Y. Asakawa, and Y. Aoyama, *Supramolec. Chem.*, **2**, 133 (1993).

25. K. Odashima, K. Yagi, K. Tohda, and Y. Umezawa, *Anal. Chem.*, **65**, 1074 (1993).

26. K. Kobayashi, M. Tominaga, Y. Asakawa, and Y. Aoyama, *Tetrahedron Lett.*, 5121 (1993).

27. R. Yanagihara and Y. Aoyama, *Tetrahedron Lett.*, 9275 (1994).

28. J. P. M. van Duynhoven, R. G. Janssen, W. Verboom, S. M. Franken, A. Casnati, A. Pochini, R. Ungaro, J. de Mendoza, P. M. Nieto, P. Prados, and D. N. Reinhoudt, *J. Am. Chem. Soc.*, **116**, 5814 (1994).

29. T. Fujimoto, R. Yanagihara, K. Kobayashi, and Y. Aoyama, *Bull. Chem. Soc. Jpn.*, **68**, 2113 (1995).

30. Y. Kikuchi and Y. Aoyama, *Bull. Chem. Soc. Jpn.*, **69**, 217 (1996).

31. Depicted by program Chem3D™ with the use of coordinates from the CSD.

32. D. J. Cram, S. Karbach, H. Kim, C. B. Knobler, E. F. Maverick, J. L. Ericson, and R. C. Helgeson, *J. Am. Chem. Soc.*, **110**, 2229 (1988); M. E. Tanner, C. B. Nobler, and D. J. Cram, *J. Org. Chem.*, **57**, 40 (1992).

33. E. Dalcanale, P. Soncini, G. Bacchilega, and F. Ugozzoli, *J. Chem. Soc., Chem. Commun.*, 500 (1989).

34. M. Vincenti, E. Dalcanale, P. Soncini, and G. Guglielmetti, *J. Am. Chem. Soc.*, **112**, 445 (1990); M. Vincenti, E. Pelizetti, E. Dalcanale, and P. Soncini, *Pure & Appl. Chem.*, **65**, 1507 (1993).

35. P. Soncini, S. Bonsignore, E. Dalcanale, and F. Ugozzoli, *J. Org. Chem.*, **57**, 4608 (1992).

36. D. A. Leigh, P. Linnane, R. G. Pritchard, and G. Jackson, *J. Chem. Soc., Chem. Commun.*, 389 (1994).

37. A. Arduini, M. Cantoni, E. Graviani, A. Pochini, A. Secchi, A. R. Sicuri, R. Ungaro, and M. Vincenti, *Tetrahedron*, **51**, 599 (1995).

38. P. Timmerman, E. A. Brinks, W. Verboom, and D. N. Reinhoudt, *J. Chem. Soc., Chem. Commun.*, 417 (1995).

39. Review: F. Diederich, *Angew. Chem. Int. Ed.*, **27**, 362 (1988); H. J. Schneider, *ibid.*, **30**, 1417 (1991).

40. J. Canceill, A. Collet, J. Gabard, F. Kotzyba-Hibert, and J. M. Lehn, *Helv. Chim. Acta*, **65**, 1894 (1982).

41. K. Odashima, A. Itai, Y. Iitaka, and K. Koga, *J. Am. Chem. Soc.*, **102**, 2504 (1980).

42. K. Odashima, A. Itai, Y. Iitaka, Y. Arata, and K. Koga, *Tetrahedron Lett.*, 4347 (1980).

43. K. Mori, K. Odashima, A. Itai, Y. Iitaka, and K. Koga, *Heterocycles*, **21**, 388 (1984).

44. K. Odashima, *Yakugaku Zassi*, **108**, 91 (1988).

45. F. Diederich, K. Dick, and D. Griebel, *Chem. Ber.*, **118**, 3588 (1985).

46. F. Diederich, K. Dick, and D. Griebel, *J. Am. Chem. Soc.*, **108**, 2273 (1986).

47. J. Canceill, L. Lacombe, and A. Collet, *J. Am. Chem. Soc.*, **108**, 4230 (1986).

48. A. V. Muehldorf, D. van Engen, J. C. Warner, and A. D. Hamilton, *J. Am. Chem. Soc.*, **110**, 6561 (1988).

49. C. A. Hunter, *J. Chem. Soc., Chem. Commun.*, 749 (1991).

50. J. E. Cochran, T. J. Parrott, B. J. Whitlock, and H. W. Whitlock, *J. Am. Chem. Soc.*, **114**, 2269 (1992).

51. W. Josten, S. Neumann, F. Vögtle, M. Nieger, K. Hagele, M. Prybylski, F. Beer, and K. Muller, *Chem. Ber.*, **127**, 2089 (1994); S. Breidenbach, J. Harren, S. Neumann, M. Nieger, K. Rissanen, and F. Vögtle, *J. Chem. Soc., Perkin 1*, 2061 (1996).

52. P. R. Ashton, D. Philp, N. Spencer, J. F. Stoddart, and D. J. Williams, *J. Chem. Soc., Chem. Commun.*, 181 (1994).

53. M. V. Reddington, A. M. Z. Slawin, N. Spencer, J. F. Stoddart, C. Vincent, and D. J. Williams, *J. Chem. Soc., Chem. Commun.*, 630 (1991). See also B. Odell, M. V. Reddington, A. M. Z. Slawin, N. Spencer, J. F. Stoddart, and D. M. Williams, *Angew. Chem. Int. Ed.*, **27**, 1547 (1988); P. R. Ashton, B. Odell, M. V. Reddington, A. M. Z. Slawin, J. F. Stoddart, and D. M. Williams, *ibid.*, **27**, 1550 (1988).

54. P. L. Anelli, P. R. Ashton, N. Spencer, A. M. Z. Slawin, J. F. Stoddart, and D. J. Williams, *Angew. Chem. Int. Ed.*, **30**, 1036 (1991).

55. D. B. Amabilino, P. R. Ashton, C. L. Brown, E. Cordova, L. A. Godinez, T. T. Goodnow, A. E. Kaifer, S. P. Newton, M. Pietraszkiewicz, D. Philp, F. M. Raymo, A. S. Reder, M. T. Rutland, A. M. Z. Slawin, N. Spencer, J. F. Stoddart, and D. J. Williams, *J. Am. Chem. Soc.*, **117**, 1271 (1995).

56. P. R. Ashton, E. J. T. Chrystal, P. T. Glink, S. Menzer, C. Schiavo, J. F. Stoddart, P. A. Taskar, and D. J. Williams, *Angew. Chem. Int. Ed.*, **34**, 1869 (1995).

57. Catenanes are irreversibly interlocked molecules and can by no means be called supramolecules. The interactions involved in this type of compound are mentioned here for convenience.

58. P. R. Ashton, T. T. Goodnow, A. E. Kaifer, D. Philp, M. M. V. Reddington, A. M. Z. Slawin, N. Spencer, J. F. Fraser, C. Vicent, and D. J. Williams, *Angew. Chem. Int. Ed.*, **28**, 1396 (1989); D. B. Amabilino, P. R. Ashton, M. S. Tolley, J. F. Stoddart, and D. J. Williams, *ibid.*, **32**, 1297 (1993); D. B. Amabilino, P. R. Ashton, J. F. Stoddart, S. Menzer, and D. J. Williams, *J. Chem. Soc., Chem. Commun.*, 2475 (1994); D. B. Amabilino, P. R. Ashton, G. R. Brown, W. Haynes, J. F. Stoddart, M. S. Tolley, J. F. Stoddart, and D. J. Williams, *ibid.*, 2479 (1994); P. R. Ashton, I. Iriepa, M. V. Reddington, N. Spencer, A. M. Z. Slawin, J. F. Stoddart, and D. J. Williams, *Tetrahedron Lett.*, 4835 (1994); P. R. Ashton, L. Perez-Garcia, J. F. Stoddart, A. J. P. White, and D. J. Williams, *Angew. Chem. Int. Ed.*, **34**, 571 (1995); P. R. Ashton, R. Bellardini, V. Balzani, A. Credi, M. T. Gandorfi, S. Menzer, L. Pérez-García, L. Prodi, J. F. Stoddart, M. Venturi, A. J. P. White, and D. J. Williams, *J. Am. Chem. Soc.*, **117**, 11171 (1995).

59. D. B. Amabilino, P. L. Anelli, P. R. Ashton, C. L. Brown, E. Córdova, L. A. Godínez, W. Hayes, A. E. Kaifer, D. Philp, A. M. Z. Slawin, N. Spencer, J. F. Stoddart, M. S. Tolley and D. J. Williams, *J. Am. Chem. Soc.*, **117**, 11142 (1995).

60. P. R. Ashton, J. A. Preece, J. F. Stoddart, M. S. Tolley, A. J. P. White, and D. J. Williams, *Synthesis*, 1343 (1994).

61. D. Armspach, P. R. Ashton, C. P. Moore, N. Spencer, J. F. Stoddart, T. J. Wear, and D. J. Williams, *Angew. Chem. Int. Ed.*, **32**, 854 (1993);

D. Armspach, P. R. Ashton, R. Ballardini, V. Balzani, A. Godi, C. P. Moore, L. Prodi, N. Spencer, J. F. Stoddart, M. S. Tolley, T. J. Wear, and D. J. Williams, *Chem. Eur. J.*, **1**, 33 (1995).

62. P. R. Ashton, J. Huff, S. Menzer, I. W. Parsons, J. A. Preece, J. F. Stoddart, M. S. Tolley, A. J. P. White, and D. J. Williams, *Chem. Eur. J.*, **2**, 123 (1996).

63. P. L. Anelli, P. R. Ashton, R. Bellardini, V. Balzani, M. Delgado, M. T. Gandorfi, T. T. Goodnow, A. E. Kaifer, D. Philp, M. Pietraszkiewicz, L. Pordi, M. V. Reddington, A. M. Z. Slawin, N. Spencer, J. F. Fraser, C. Vicent, and D. J. Williams, *J. Am. Chem. Soc.*, **114**, 193 (1992).

CHAPTER 10

INTERLIGAND INTERACTIONS
IN COORDINATION CHEMISTRY

In this chapter we examine the possible involvement of a CH/π interaction in the chemistry of metal complexes. In these compounds, ligands are arranged so as to be proximate to each other; this is mediated by the coordination of the central metal ion. This may give the weak intramolecular forces a chance to play both cooperative and effective roles in these compounds.

10.1. CRYSTAL STRUCTURES OF COORDINATION COMPOUNDS

Short interligand C/C or CH/C distances are found among the X-ray crystallographic data of coordination compounds (Table 10.1). Figure 10.1 illustrates a result.

10.2. CONFORMATION OF COORDINATION COMPOUNDS

Sigel extensively studied the conformations of a number of ternary metal complexes bearing aromatic moieties at both termini of the ligands in solution. These include complexes with biologically important molecules such as adenosine triphosphate (ATP) and tryptophane (Trp). Figure 10.2 illustrates the conformations of metal complexes of Trp and ATP,[11] and of ATP and 1,10-phenanthroline

TABLE 10.1. X-ray Crystallographic Data of Coordination Compounds

[Cu(phen)(phenylpropionate)] (phen = 1,10-phenanthroline) **1**[1]
[Rh(L)(NH$_3$)$_2$ · dibenzo-3n-crown-n ether] [PF$_6$]
(L = cycloocta-1,5-diene or norbornadiene)[2]
[Cu(Ha)(Phe)] (Ha = histamine, Phe = L-phenylalanine) **2**[3]
Bis[*N*-(*R*)-1-phenylethylsalicylideneiminato]-zinc **3**[4]
[η^5-CH$_3$C$_5$H$_4$)Mn(CO)(1,1'-bis(diphenylphosphino)ferrocene)] **4**[5]
[Co(pyridoxy-L-phenylalaninato)$_2$]^6
[LMo(CO)$_2$ (η^3-C$_6$H$_5$CH=CHCH$_2$)] [L = bis(3,5-dimethyl-1-pyrazolyl)-
   phosphinate] **5**[7]
2-Pyridyl-1,1'-ferrocenedicarboximide **6**[8]
[Ru(4-methylpyrimidine-2-thione)(bipy)$_2$] (bipy = bipyridyl) **7**[9]
[Co(*S*-α-methyltyrosine)(2*R*,5*R*,8*R*,11*R*-2,5,8,11-tetraethyl-1,4,7,10-
   tetraazacyclododecane)]^{10} (Fig. 10.1)

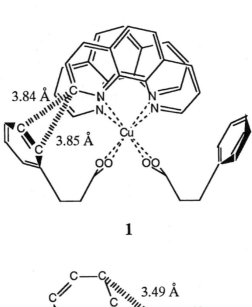

1

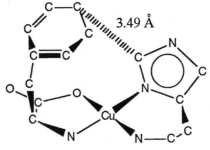

2

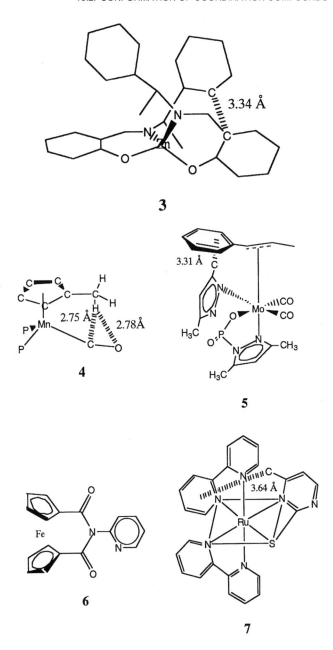

3

4

5

6

7

(phen),[12] existing in solution as deduced from NMR spectroscopy and potentiometric titrations. Folded conformations containing aromatic/aromatic stacking generally occur in complexes with aromatic ligands.[13]

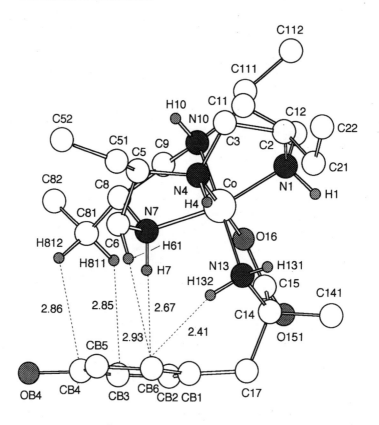

Figure 10.1. Perspective view of [Co(S-α-methyltyrosine)(2R,5R,8R,11R)-2,5,8,11-tetraethyl-1,4,7,10-tetraazacyclododecane].

They also observed alkyl/aromatic interactions in a number of coordination compounds. In ternary complexes such as M(phen)(aa), M(bipy)(aa),[14] M(ATP)(aa)[15] [M = Cu^{2+} or Zn^{2+}, bipy = 2,2′-bipyridyl, aa = R-CH(NH$_2$)COOH: alaninate, aminobutyrate, norleucinate, leucinate, isoleucinate], and Zn(phen)(aca) [aca = Me$_2$CH-(CH$_2$)$_n$COO, n = 1–4],[16] the alkyl groups are curled over the aromatic rings to form folded rather than extended conformation (Fig. 10.3).

Table 10.2 lists differences in the complexation-induced NMR chemical shift (CIS) and percentages of the folded conformations, determined by potentiometric pH titrations, for zinc complexes with a series of isoalkanecarboxylates, Zn(phen)(aca).

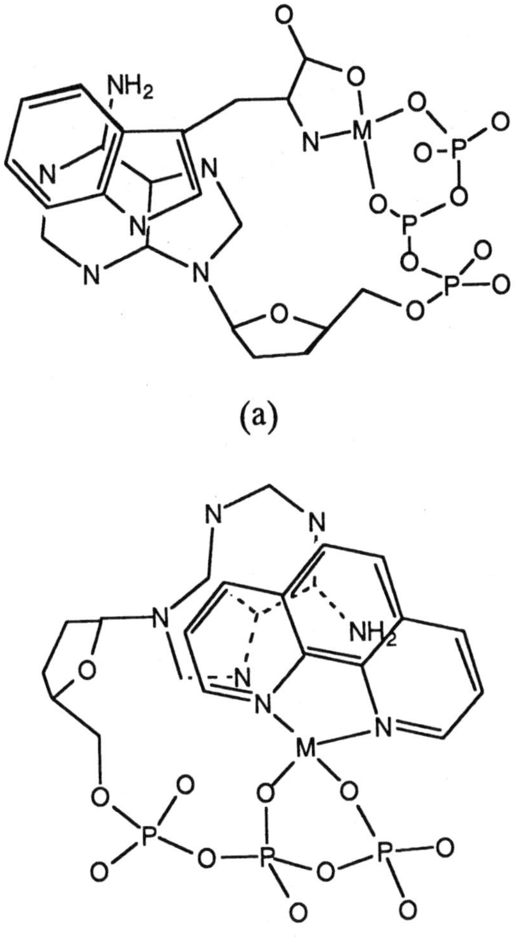

(a)

(b)

Figure 10.2. Solution conformations of metal complexes: (a) M(ATP)(Trp);
(b) M(ATP)(phen).

The percentage of the folded species increased according to the size
of the alkyl moiety. Maximum values of both CIS and the percentage
for the closed conformers were reached at 6-methylheptanoate
($n = 4$). The proportion of the folded conformation was found to be
influenced by solvent polarity: addition of ethanol or dioxane to an
aqueous solution increased the proportion of the folded species.

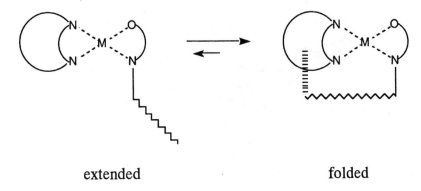

extended   folded

Figure 10.3. Extended and folded conformations of a ternary metal complex.

This finding is remarkable and contrary to simple hydrophobicity consideration.

Inoue et al. reported on an unusual conformational stability of sterically crowded atropisomerism of porphyrin derivatives.[17]

8

5,10,15,20-Tetrakis(2′-biphenyl)porphyrin **8a** ($R^1 = R^3 = Ph, R^2 = R^4 = H$) has four atropisomers denoted as α^4, $\alpha^3\beta$, $\alpha\alpha\beta\beta$, and $\alpha\beta\alpha\beta$ (Fig. 10.4). When $AlMe_3$ is treated with the α^4 isomer, the resulting methylaluminium porphyrin α^4 should have two isomers, α–α^4 and β–α^4, with different degrees of steric crowding.

TABLE 10.2. Complexation-Induced NMR Chemical Shift and Percentages of the Folded Conformations for a Series of Zinc Complexes

		Zn(phen)(aca) folded (%)		
n	CIS^a	H_2O	50% Ethanol	50% Dioxane
0	0.20	2	19	28
1	0.30	19	21	26
2	0.46	19	37	37
3	0.54	22	41	35
4	0.54	32	41	37

$^a \delta Zn(aca) - \delta Zn(phen)(aca)$ in D_2O (ppm). aca $= Me_2CH(CH_2)_nCOO$, phen $= 1,10$-phenanthroline

The β isomer showed a methyl signal at a high upfield ($\delta - 5.89$) primarily due to shielding by the porphyrin ring. However, the degree of the upfield shift was more remarkable ($\delta - 6.92$) in the α isomer. The additional shielding effect is considered to be brought about by the four neighboring biphenyl groups. When the sterically less crowded $\beta - \alpha^4$ was allowed to stand in C_6D_6 at 35°C, the population of the starting isomer decreased gradually to give the more crowded atropisomers. The $\beta - \alpha^4$ isomer remained only 68, 17, and 0% after 1, 6, and 8 hours, respectively. On the other hand, the sterically crowded $\alpha - \alpha^4$ isomer underwent a much slower isomerization under identical conditions: the population of the starting material was still at 40% even after 50 hours. Thus, the proximal orientation of the 2'-biphenyl-group to the CH_3-Al moiety was shown to be much preferred over the distal orientation.

A simpler homolog carrying only one biphenyl group, methyl-[5,10,15-triphenyl-20-(2'-biphenyl)porphyrinato]aluminium **8b** ($R^1 =$ Ph, $R^2 = R^3 = R^4 = H$), was prepared. A solution of the reaction mixture with $AlMe_3$ shows two NMR singlets corresponding to the Me-Al of the α and β atropisomers in a ratio of $\alpha:\beta = 64:36$. The preference of the proximal CH_3/aromatic isomer increased when an electron donating methoxy group was introduced on the 2'-phenyl group **8c** ($R^1 = C_6H_4OCH_3$-p, $R^2 = R^3 = R^4 = H$, $\alpha:\beta = 72:28$). The reverse was true for the introduction of an electron-withdrawing CF_3 group **8d** ($R^1 = C_6H_4CF_3$-p, $R^2 = R^3 = R^4 = H$, $\alpha:\beta = 57:43$).

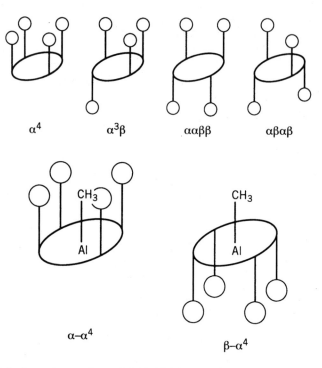

Figure 10.4. Atropisomerism of 5,10,15,20-tetrakis(2'-biphenyl)porphyrin **8**.

In light of these findings, the importance of the CH/π interaction in determining the stability of these porphyrins was suggested.

10.3. SELECTIVITY IN THE FORMATION OF COORDINATION COMPOUNDS

Selective formation of only one of the stereoisomers is often noted in coordination chemistry. Careful examination revealed that attractive interaction between an aliphatic and an aromatic part in the ligands always exists whenever significant selectivity was observed.[18] For example, Okawa found that an isomer (Λ-cis β_1; four stereoisomers are possible for these compounds due to the coordination attitude of l-menthyloxy-3-benzoylacetone and the asymmetry about the central metal) was preferentially produced in mixed chelate compounds, [Co(SB)(1-moba)] (SB = N,N'-disalicylideneethylenediamine), moba = R = (1-l-menthyloxy-3-benzoylacetone).[19] They attributed the

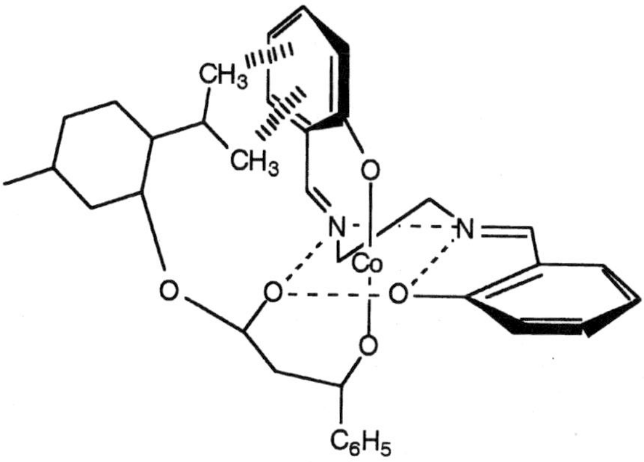

Figure 10.5. Suggested interactions in a mixed-chelate compound, [Co(SB)-(1-moba)] (Λ-cis β_1 isomer).

TABLE 10.3. Thermodynamic Parameters for the Formation of Tetrakis(2,4,6-trialkylphenyl)porphyrinato-iron(III) 9^a

R^1	R^2	ΔH (kcal mol^{-1})	ΔS (e.u.)	ΔG (kcal mol^{-1}) ($-35°$C)
H	H	-26.5	-75	-8.3
H	Me	-20.6 (-5.9)^b	-73	-3.2
Me	H	-28.0	-76	-9.9
Me	Me	-27.8 (-0.2)	-86	-7.3
Et	H	-27.2	-69	-10.8
Et	Me	-28.2 (1.0)	-87	-7.5
Pr^i	H	-27.2	-73	-9.8
Pr^i	Me	-29.1 (1.9)	-91	-7.4

^{a} In CDCl$_3$.
^{b} Data in parentheses are the differences in enthalpy between 1-methyl (R^2 = H) and 1,2-dimethyl (R^2 = Me) compounds ($\Delta\Delta H$).

stereo-selectivity to the interligand CH/π interaction, which may only occur in this stereoisomer, between the *l*-menthyl group and an aromatic ring of the other ligand (Fig. 10.5).

Nakamura and Nakamura studied the thermodynamics of complex formation of imidazole with low-spin tetrakis(2,4,6-trialkylphenyl)-porphyrinato-iron(III) **9**.[20] The results are listed in Table 10.3.

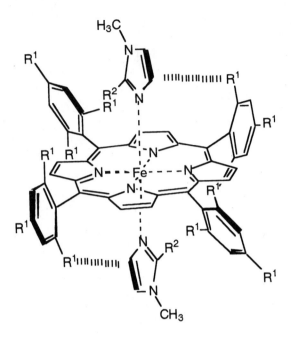

Figure 10.6. Suggested interactions in the imidazole complex of tetrakis-(2,4,6-trialkylphenyl)porphyrinato-iron(III) **9**.

In contrast to the lower-alkyl analogs (R^1 = H and Me), triethyl-phenyl and triisopropylphenyl derivatives (R^1 = Et and Pr^i) showed a larger negative enthalpy of formation with 1,2-dimethylimidazole (R^2 = Me) than with 1-methylimidazole (R^2 = H) as the axially co-ordinated ligand. The difference in the enthalpy of formation be-tween R^2 = H and R^2 = Me changes from -5.9 for R^1 = H, -0.2 (R^1 = Me), 1.0 (R^1 = Et), and 1.9 kcal mol^{-1} (R^1 = Pr^i). Thus, the enthalpy difference between **9** (R^1 = Pr^i, R^2 = Me) and **9** (R^1 = H, R^2 = Me) amounts to 7.8 [$= 1.9 - (-5.9)$] kcal mol^{-1}. Although this value seems to be too large to be solely ascribed to CH/π interaction and should include other forces such as the deformation of a porphyrin ring,[21] the results nonetheless demonstrate the importance of an attractive interaction of the alkyl group with the imidazole moiety (Fig. 10.6).

Imai et al. compared the complexation of a series of porphyrins **10–12** with a variety of amine ligands.[22] These porphyrin hosts vary in their degree of preorganization. Host **10** bears a "bis-roof" structure and is more effectively preorganized for complexation than **11** or **12**. Table 10.4 summarizes the results.

10: R^1–R^2 = $(CH_2)_5$, R^3–R^4 = $(CH_2)_5$, 11: R^1 = R^2 = CH_3,

R^3–R^4 = $(CH_2)_5$, 12: R^1 = R^2 = R^3 = R^4 = CH_3

Secondary amines such as azetidine, pyrrolidine, and piperidine, which fit well in the cavity, are bound effectively to the porphyrin hosts. Host **10**, with its "bis-roof" structure, binds most effectively azetidine (**a**), pyrrolidine (**b**), and diethylamine (**d**), while the bulkier piperidine (**c**) and aromatic bases such as pyridine or isoquinoline (not listed in Table 10.4) are more selectively bound to host **11**. This is reasonable in view of steric considerations. The best selectivity was noted with host **10**, where the specificity of recognition for **a** (4-membered ring) versus **c** (6-membered ring) is 64:1. Such remarkable selectivity of binding was not found for **11** (2.8:1) or **12** (3:1). The above binding behaviors of the guest amines was interpreted by these authors as an attractive force between CHs and the host aromatic groups (Fig. 10.7).

TABLE 10.4. Binding Constants[a] of Zinc Porphyrins 10–12 with Various Amines

	a	b	c	d	e
10	1.4×10^7	2.7×10^6	2.2×10^5	5.8×10^4	7.7×10^4
11	1.8×10^6	1.3×10^6	6.5×10^5	1.2×10^4	7.6×10^4
12	7.5×10^5	6.9×10^5	2.5×10^5	4.2×10^3	4.9×10^4

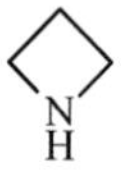

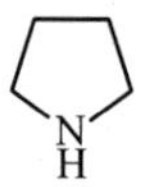

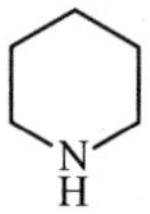

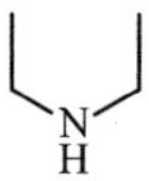

[a] At 298 K in $CHCl_3$ dm^3 mol^{-1}.

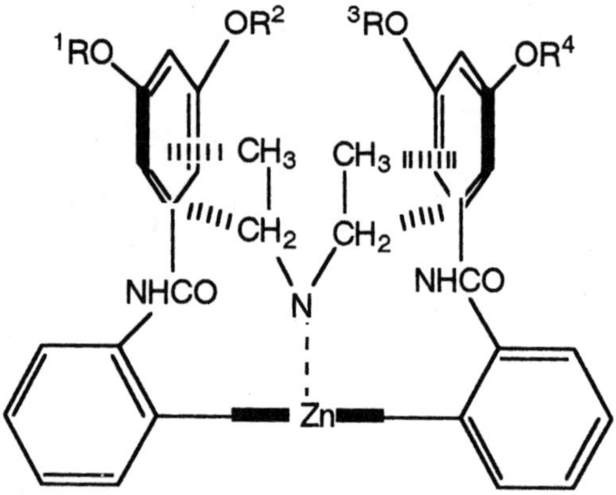

Figure 10.7. Interaction of a secondary amine with porphyrin hosts **10–12**.

Yamanari et al. studied the stereochemistry of formation for cobalt complexes with pyrimidine-2-thionates.[23] Selective production of one of the linkage isomers with facial geometry was only recorded when a methyl group was introduced into the pyrimidine ring. An analogous conclusion has been obtained for ruthenium complexes.[9,24] They attributed the results as consequences of the CH/π interaction (Fig. 10.8).

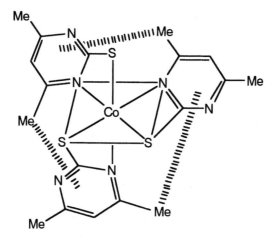

Figure 10.8.

REFERENCES

1. E. Dubler, U. K. Haring, K. H. Scheller, P. Baltzer, and H. Sigel, *Inorg. Chem.*, **23**, 3785 (1984).
2. H. M. Colquhoun, S. M. Doughty, J. F. Stoddart, A. M. Slawin, and D. J. Williams, *J. Chem. Soc., Dalton*, 1639 (1986); H. M. Colquhoun, F. Stoddart, and D. J. Williams, *Angew. Chem. Int. Ed.*, **25**, 487 (1986).
3. O. Yamauchi, A. Odani, T. Kohzuma, H. Masuda, K. Toriumi, and K. Saito, *Inorg. Chem.*, **28**, 4066 (1989).
4. H. Sakiyama, H. Okawa, N. Matsumoto, and S. Kida, *J. Chem. Soc., Dalton*, 2935 (1990).
5. S. Onaka, H. Furuta, and S. Takagi, *Angew. Chem. Int. Ed.*, **32**, 87 (1993).
6. K. Jitsukawa, K. Iwai, H. Masuda, H. Ogoshi, and H. Einaga, *Chem. Lett.*, 303 (1994).
7. S. K. Chowdhury, V. S. Joshi, A. G. Samuel, V. G. Puranik, S. S. Tavale, and A. Sarkar, *Organometallics*, **13**, 4092 (1994).
8. T. Moriuchi, I. Ikeda, and T. Hirao, *Organometallics*, **14**, 3578 (1995). An intermolecular CH/π contact (2.97 Å) was found between the cyclopentadienyl hydrogen and pyridyl ring.
9. K. Yamanari, T. Nozaki, A. Fuyuhiro, Y. Kushi, and S. Kaizaki, *J. Chem. Soc., Dalton*, 2851 (1996).
10. S. Tsuboyama et al., unpublished data. We thank Dr. Sei Tsuboyama for information.

11. H. Sigel and C. F. Naumann, *J. Am. Chem. Soc.*, **98**, 730 (1976); P. R. Mitchell, B. Prijs, and H. Sigel, *Helv. Chim. Acta*, **62**, 1723 (1979).

12. P. R. Mitchell and H. Sigel, *J. Am. Chem. Soc.*, **100**, 1564 (1978).

13. Review: H. Sigel, *Angew. Chem. Int. Ed.*, **14**, 394 (1975); H. Sigel, *Pure Appl. Chem.*, **61**, 923 (1989).

14. B. E. Fischer and H. Sigel, *J. Am. Chem. Soc.*, **102**, 2998 (1980).

15. H. Sigel, B. E. Fischer, and E. Farkas, *Inorg. Chem.*, **22**, 925 (1983).

16. G. Liang, R. Tribolet, and H. Sigel, *Inorg. Chem.*, **27**, 2877 (1988).

17. H. Sugimoto, T. Aida, and S. Inoue, *J. Chem. Soc., Chem. Commun.*, 1411 (1995).

18. Review: H. Okawa, *Coord. Chem. Rev.*, **92**, 1 (1988). H. Okawa, Y. Numata, A. Mio, and S. Kida, *Bull. Chem. Soc. Jpn.*, **53**, 2248 (1980); M. Nakamura, H. Okawa, and S. Kida, *Chem. Lett.*, 547 (1981); H. Okawa, K. Ueda, and S. Kida, *Inorg. Chem.*, **21**, 1594 (1982); M. Nakamura, H. Okawa, and S. Kida, *Inorg. Chim. Acta*, **96**, 111 (1984); M. Nakamura, H. Okawa, and S. Kida, *Bull. Chem. Soc. Jpn.*, **58**, 3377 (1985); M. Nakamura, H. Okawa, T. Ito, M. Kato, and S. Kida, *ibid.*, **60**, 539 (1987); S. Maeda, M. Nakamura, H. Okawa, and S. Kida, *Polyhedron*, **6**, 583 (1987).

19. M. Nakamura, H. Okawa, T. Inazu, and S. Kida, *Bull. Chem. Soc. Jpn.*, **55**, 2400 (1982).

20. M. Nakamura and N. Nakamura, *Chem. Lett.*, 181 (1990); M. Nakamura, *Inorg. Chim. Acta*, **161**, 73 (1989).

21. M. Nakamura, *Bull. Chem. Soc. Jpn.*, **68**, 197 (1995).

22. H. Imai and Y. Uemori, *J. Chem. Soc., Perkin 2*, 1793 (1994); H. Imai, S. Nakagawa, and E. Kyuno, *J. Am. Chem. Soc.*, **114**, 6719 (1992). See also H. Imai and E. Kyuno, *Inorg. Chem.*, **29**, 2416 (1990).

23. K. Yamanari, S. Dogi, K. Okusato, T. Fujihara, A. Fuyuhiro, and S. Kaizaki, *Bull. Chem. Soc. Jpn.*, **67**, 3004 (1994).

24. K. Yamanari, T. Nozaki, A. Fuyuhiro, and S. Kaizaki, *Chem. Lett.*, 35 (1996).

CHAPTER 11

SPECIFIC INTERACTIONS IN PROTEIN STRUCTURES

In view of the ubiquitous presence of CH groups in nature, it is intriguing to examine the possibility of the involvement of this weak attractive force in biochemistry, since π-containing groups are also abundant in biochemically important molecules. These include the side-chain groups of aromatic amino acids (tryptophane, tyrosine, phenylalanine, and histidine), the aromatic rings of nucleic acid bases, porphyrins, nicotineamide, and flavins. Tertiary structures of proteins relevant to cell biology are currently being elucidated at the atomic level by means of X-ray crystallography and NMR spectroscopy. The molecular forces invoked should therefore be analyzed with the same kind of precision.

11.1. GENERAL CONSIDERATIONS

In the late 1970s, the authors suggested the potential role of the CH/π interaction in protein structures such as enzymes, hemoglobins, and immunoglobulins.[1] Several papers have since appeared which comment upon the occurrence of CH/π interaction in protein structures.

Kim et al.[2] reported on the crystal structure of carboxypeptidase A covalently modified by an inactivator 2-benzyl-3,4-epoxybutanoic acid. The phenyl group of the inhibitor contacts with the enzyme at Leu203 and Ile243 in a perpendicular fashion on each side of the ring, the distances being 3.9 and 3.5 Å, respectively. They argued that this was a consequence of CH/π interaction.

Iwasaki et al.[3] studied the inhibitory effect of cerulenin analogs on yeast fatty acid synthase. Cerulenin (R = CH$_3$) is a potent inhibitor of fatty acid synthase since the (E,E)-1,4-diene system of the side-chain structure is found to be essential in effective binding to the enzyme.

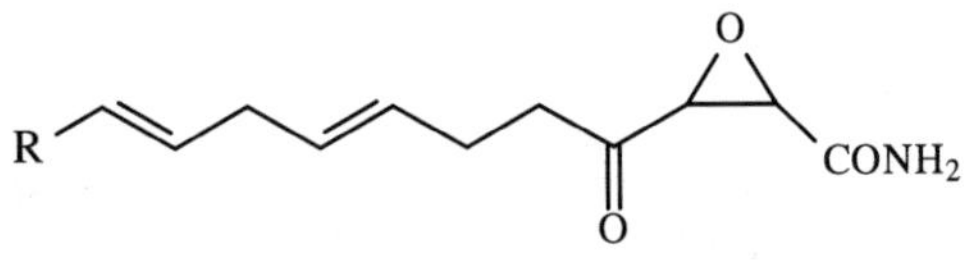

cerulenin (R = Me)

Derivatives (R = H, n-C$_3$H$_7$) bearing the (E,E)-1,4-diene moiety retained this activity, while tetrahydrocerulenin was found to be inactive. Truncation or elongation of the chain length appreciably decreased the inhibitory effect. They speculated that an attractive interaction, such as a CH/π interaction, is playing a role between the (E,E)-Δ7,10-double bonds and the side-chain groups of the protein.

Hirama et al. determined the solution structure of an antitumor antibiotic neocarzinostatin by NMR: a complex of an unstable chromophore **1** and an apoprotein of low molecular weight (113 residues).[4]

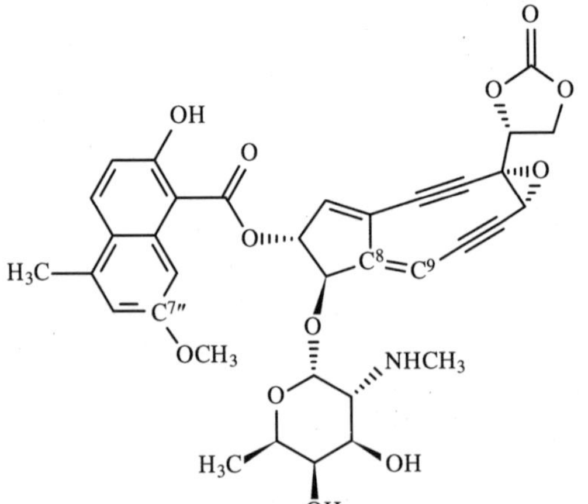

1

Interatomic distances determined by NOE measurements were 3.8, 3.5, and 3.4 Å, respectively, for C^8/Leu45(Cδ2), $C^{7''}O\underline{C}$/Phe52(Cγ), and $C^{7''}O\underline{C}$/Phe52(Cζ). The Leu45 side chain in the apoprotein was found close to the C^8–C^9 double bond of the enediyne core. An upfield shift of about 1 ppm observed for the C(7'')OC$\underline{H}_3$ resonance of **1** upon complexation was attributed to the diamagnetic anisotropy of the nearby aromatic ring of Phe52. The findings were followed by a crystallographic study by Kim et al.[5] using the same neocarzinostatin complex; the aromatic rings of Phe52 and Phe78 were shown to be close and perpendicular to the enediyne chromophore.

2   **3**

The complexing abilities of the apoprotein were determined for a series of synthetic model compounds.[6] The binding constant of a demethylated analog (**3**: K_a 4.2×10^3 mol^{-1}) was approximately one-third of that of **2** (K_a 1.3×10^4 mol^{-1}). The difference in the binding energy was estimated to be about 0.7 kcal mol^{-1}, which corresponds to the loss of a CH/π interaction between OCH$_3$ and the aromatic ring of Phe52.

Based on the findings of the interligand interactions involved in a number of cobalt(III) complexes bearing a pyridoxal-5-phosphate as a ligand, Masuda et al. commented upon the possible role of the CH/π interaction in biochemistry.[7]

pyridoxal-5-phosphate

The biochemical significance of certain functional groups (3-OH, etc.) on the pyridoxal ring has been known. However, the role of the 2-methyl group remains to be clarified, in view of the finding that the displacement of this group abolishes the biological activity of the coenzyme. The X-ray structure of aspartate aminotransferase bound with its cofactor has demonstrated that the aromatic rings of tyrosine and tryptophane in the enzyme are close to the 2-methyl group in the pyridoxal pyridine ring.[8]

Chakrabarti and Samanta examined the possibility of a CH/π interaction by analyzing the structures of proteins.[9] They set about surveying the crystal data of enzymes (dehydrogenases, reductases, kinases, etc.) complexed with cofactors bearing an adenine moiety: NAD, NADP, FAD, AMP, ADP, and ATP. Carbon/nitrogen and carbon/carbon distances around 3.7 Å, between side-chain groups and an adenine ring, were collected. A number of short C/N_{sp^2} and C/C_{sp^2} distances were found, in 19 protein/cofactor complexes, between adenine aromatic ring and carbons in the side-chain groups. Residues bearing branched aliphatic groups such as valine, leucine, and isoleucine were often found, but it is remarkable that lysine and arginine[10] are also involved in this type of complex; there they noted a significant overlap involving the nonpolar part of the side-chain in such a way that a carbon atom is in the shortest contact with an adenine nitrogen atom. From this they invoked an important role of the CH/N_{sp^2} interaction in specific recognition of the cofactors with an adenine moiety. The findings are consistent with the discussions in the following sections.

11.2. EXPLORING THE POSSIBILITY OF CH/π INTERACTIONS IN PROTEIN CRYSTAL DATA

A program (CHPI) was written in order to search short contacts between CH groups and π systems in protein structures registered in Brookhaven Protein Data Bank (PDB).[11] Figure 11.1 illustrates the method. The π-system may be an aromatic group (five- or six-membered, or fused) or a double bond (C=C, C=O, or C=N). The hydrogen may be a part of an alkyl group (CH_3, CH_2, or CH), a CH in aromatic rings, N^+H_3, NH_2, NH, OH, or SH group.

To participate in a XH/π interaction, a hydrogen should be positioned above the π plane, most preferably above the sp^2 atom (region 1 in Fig. 11.1). To cover other possibilities, several kinds of H/X_{sp^2} distance (D_{atm}, D_{pln}, and D_{lin}) and angle parameters (θ and ω) were defined.

Table 11.1 is an output from the program CHPI analyzing the protein, bovine pancreatic trypsin inhibitor (BPTI),[12] using hydrogen

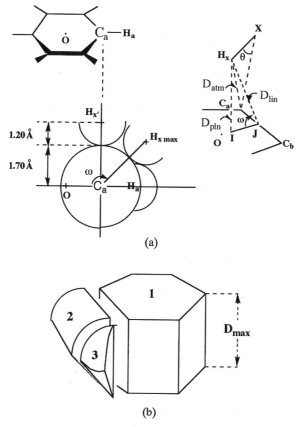

Figure 11.1. Method for exploring XH/π contacts (X = C, N, O, or S). An example is given for a six-membered π-system. (*a*) O: center of the π-plane. C_a and C_b: nearest and second nearest sp^2 atoms, respectively, to the H_X hydrogen. ω: dihedral angle defined by C_aOC_b and $H_XC_aC_b$ planes. θ: H_X–X–C_a angle. D_{pln}: perpendicular distance between H_x and the π-plane (H_X/I). D_{atm}: H_X/C_a interatomic distance. D_{lin}: distance between H_X and the line C_a–C_b (H_X/J). (*b*) Regions to be searched. Region 1: zone where H_X is above the ring. Regions 2 and 3: zones where H_x is out of region 1 but may interact with π-orbitals. The program was run to search for short H_X/π contacts with the following conditions: D_{max} = 3.05 Å: (1.2 Å + 1.7 Å) × 1.05; $D_{pln} < D_{max}$ (region 1); $D_{lin} < D_{max}$ (region 2); $D_{atm} < D_{max}$ (region 3); ω_{max} = 127.5°, $-\omega_{max} < \omega < \omega_{max}$; $\theta < 62.2°$.

(deuterium) coordinates from a neutron diffraction study. Crystal data in PDB do not necessarily contain the coordinates of hydrogen atoms. In such cases, hydrogens were generated on these nonhydrogen atoms and their positions were optimized.[13] Table 11.2 presents the results

TABLE 11.1. Computer Output from a CHPI Analysis of BPTI (Neutron Data)

	RES	I	VPI	1	2	3	4	5	6
PRTN	HIS	1	FIV	CG	ND1	CE1	NE2	CD2	
PRTN	PHE	1	SIX	CG	CD1	CE1	CZ	CE2	CD2
PRTN	TYR	1	SIX	CG	CD1	CE1	CZ	CE2	CD2
PRTN	TRP	1	FIV	CG	CD1	NE1	CE2	CD2	
PRTN	TRP	2	SIX	CE2	CD2	CE3	CZ3	CH2	CZ2

```
RANGE -127.500  < OMEGA < 127.500
RANGE    0.000  < THETA <  62.200
RANGE    0.000  < Dmax  <   3.050
```

pi					HX			geometry					
ID	RES	I	VPI VATM	N	ID	RES VATM	N	DATM	DPLN	DLIN	OMEGA	THETA	RG
4	PHE	1	SIX CE2	5	2	PRO HB	9	2.79	2.73	[2.74]	93.15	49.93	2
4	PHE	1	SIX CE2	5	2	PRO HB	10	2.93	2.40	[2.93]	124.99	59.35	2
4	PHE	1	SIX CG	1	42	ARG HB	14	3.02	2.79	[2.99]	111.07	50.64	2
4	PHE	1	SIX CD1	2	42	ARG HB	15	2.68	[2.66]	2.68	83.06	33.98	1
4	PHE	1	SIX CD2	6	42	ARG HD	18	[3.00]	2.96	****	98.95	18.85	3
10	TYR	1	SIX CZ	4	12	GLY HA	6	2.59	2.52	[2.55]	99.17	12.43	2
10	TYR	1	SIX CD2	6	41	LYS HG	15	[2.81]	2.46	****	118.89	30.11	3
10	TYR	1	SIX CG	1	41	LYS HD	16	[2.88]	2.71	****	109.73	40.83	3
10	TYR	1	SIX CE2	5	41	LYS HE	18	2.58	[2.45]	2.49	80.10	17.06	1
21	TYR	1	SIX CZ	4	48	ALA HB	10	2.62	[2.60]	2.61	84.91	3.06	1
22	PHE	1	SIX CG	1	9	PRO HB	10	[2.64]	2.64	****	92.32	16.32	3
22	PHE	1	SIX CE1	3	9	PRO HD	14	2.73	2.47	[2.71]	114.36	27.21	2
22	PHE	1	SIX CE1	3	24	ASN HB	11	2.87	2.58	[2.84]	114.78	28.78	2
22	PHE	1	SIX CE2	5	31	GLN HB	12	[2.88]	2.69	****	111.13	23.52	3
23	TYR	1	SIX CZ	4	1	ARG DH1	24	3.05	2.68	[3.00]	116.60	56.14	2
23	TYR	1	SIX CE1	3	25	ALA HA	7	[2.59]	2.40	****	111.95	29.86	3
23	TYR	1	SIX CG	1	55	CYS HB	10	2.84	[2.79]	2.79	88.88	45.80	1
33	PHE	1	SIX CD1	2	9	PRO HB	9	2.74	[2.70]	2.71	85.59	9.35	1
33	PHE	1	SIX CE2	5	20	ARG HB	14	[2.83]	2.79	****	98.87	37.71	3
33	PHE	1	SIX CE2	5	22	PHE HB	15	2.86	2.28	[2.79]	125.11	3.21	2
33	PHE	1	SIX CE1	3	35	TYR HB	15	2.99	2.70	[2.95]	113.90	17.26	2
35	TYR	1	SIX CD1	2	37	GLY H	5	2.80	[2.55]	2.73	69.09	13.14	1
35	TYR	1	SIX CE2	5	37	GLY HA	6	[3.00]	2.93	****	101.77	46.59	3
35	TYR	1	SIX CZ	4	40	ALA HB	10	2.66	2.49	[2.66]	110.46	15.75	2
35	TYR	1	SIX CG	1	44	ASN DD2	13	2.65	[2.59]	2.64	79.91	26.81	1
45	PHE	1	SIX CG	1	51	CYS HA	8	2.85	[2.61]	2.79	69.19	39.43	1
45	PHE	1	SIX CG	1	51	CYS HB	9	2.92	2.54	[2.86]	117.39	41.41	2

Number of H/π interactions: 27

obtained by the use of the generated (and optimized) hydrogen coordinates for BPTI. Interactions satisfying the following conditions were collected. $D_{max} = 3.05$ Å, $\theta < 62.2°$, $\omega < 127.5°$. Criteria for detecting CH/π bonds were as follows. $D_{pln} < D_{max}$ for hydrogens in region 1. $D_{lin} < D_{max}$ and $D_{atm} < D_{max}$ for those falling within regions 2 and 3, respectively. The numbers in the brackets in the tables correspond to these values.

TABLE 11.2. Computer Output from a CHPI Analysis of BPTI (Generated H)

	RES	I	VPI	1	2	3	4	5	6
PRTN	HIS	1	FIV	CG	ND1	CE1	NE2	CD2	
PRTN	PHE	1	SIX	CG	CD1	CE1	CZ	CE2	CD2
PRTN	TYR	1	SIX	CG	CD1	CE1	CZ	CE2	CD2
PRTN	TRP	1	FIV	CG	CD1	NE1	CE2	CD2	
PRTN	TRP	2	SIX	CE2	CD2	CE3	CZ3	CH2	CZ2

```
RANGE  -127.500  < OMEGA < 127.500
RANGE     0.000  < THETA <  62.200
RANGE     0.000  < Dmax  <   3.050
```

pi						HX				geometry					
ID	RES	I	VPI	VATM	N	ID	RES	VATM	N	DATM	DPLN	DLIN	OMEGA	THETA	RG
4	PHE	1	SIX	CE2	5	2	PRO	HB	8	2.80	2.75	[2.75]	91.20	52.81	2
4	PHE	1	SIX	CD1	2	42	ARG	HB	8	2.87	[2.83]	2.86	82.70	46.25	1
4	PHE	1	SIX	CG	1	42	ARG	HB	9	3.06	2.83	[3.03]	110.96	52.20	2
10	TYR	1	SIX	CZ	4	12	GLY	HA	5	2.64	2.59	[2.62]	99.36	20.40	2
10	TYR	1	SIX	CD2	6	41	LYS	HG	11	[2.83]	2.48	****	118.82	30.62	3
10	TYR	1	SIX	CG	1	41	LYS	HD	15	[3.01]	2.79	****	112.38	48.20	3
10	TYR	1	SIX	CE2	5	41	LYS	HE	18	2.54	[2.36]	2.45	74.74	5.55	1
21	TYR	1	SIX	CZ	4	48	ALA	HB	10	2.64	[2.63]	2.64	86.31	7.80	1
22	PHE	1	SIX	CG	1	9	PRO	HB	7	2.57	2.56	[2.57]	96.82	16.19	2
22	PHE	1	SIX	CZ	4	9	PRO	HG	10	3.03	[2.98]	3.02	80.74	29.30	1
22	PHE	1	SIX	CE1	3	9	PRO	HD	14	2.70	2.42	[2.67]	114.87	27.46	2
22	PHE	1	SIX	CE1	3	24	ASN	HB	9	3.02	2.74	[2.98]	113.02	38.45	2
22	PHE	1	SIX	CE2	5	31	GLN	HB	9	[2.90]	2.70	****	110.98	23.77	3
23	TYR	1	SIX	CE1	3	25	ALA	HA	4	[2.63]	2.38	****	115.25	33.18	3
23	TYR	1	SIX	CZ	4	28	GLY	HA	5	[2.92]	2.36	****	126.22	23.20	3
23	TYR	1	SIX	CG	1	55	CYS	HB	8	2.83	2.78	[2.78]	91.83	44.79	2
23	TYR	1	SIX	CE2	5	56	GLY	HA	5	2.94	2.53	[2.94]	120.84	29.19	2
33	PHE	1	SIX	CD1	2	9	PRO	HB	8	2.72	[2.68]	2.69	87.08	8.33	1
33	PHE	1	SIX	CE2	5	20	ARG	HB	9	[2.84]	2.80	****	99.22	38.14	3
33	PHE	1	SIX	CE2	5	22	PHE	HB	8	2.93	2.43	[2.88]	122.61	10.46	2
33	PHE	1	SIX	CZ	4	35	TYR	HB	9	3.08	2.80	[3.02]	112.01	12.39	2
35	TYR	1	SIX	CD1	2	36	GLY	H	2	[3.00]	2.54	****	122.34	59.26	3
35	TYR	1	SIX	CD1	2	37	GLY	H	2	2.82	[2.60]	2.77	70.09	11.42	1
35	TYR	1	SIX	CE2	5	37	GLY	HA	5	[2.91]	2.85	****	101.57	40.87	3
35	TYR	1	SIX	CZ	4	40	ALA	HB	10	[2.71]	2.51	****	112.19	20.81	3
35	TYR	1	SIX	CG	1	44	ASN	HD2	14	2.63	[2.58]	2.60	81.66	21.20	1
45	PHE	1	SIX	CE1	3	43	ASN	HA	4	[3.01]	2.42	****	126.56	27.43	3
45	PHE	1	SIX	CG	1	51	CYS	HA	4	2.90	[2.63]	2.83	68.33	42.71	1
45	PHE	1	SIX	CG	1	51	CYS	HB	9	2.86	2.51	[2.81]	116.66	38.37	2

Number of H/π interactions: 29

11.3. INTERACTIONS IN HEMOGLOBIN

The heme group in hemoglobins and myoglobin is a large π-system surrounded by a number of nonpolar residues. To see whether there is evidence for the CH/π interaction, a CHPI analysis was carried out. Hydrogen atoms were generated and their positions optimized before the analysis, as mentioned above. H/C_{sp^2} distances shorter than the sum of the van der Waals distances [(1.2 Å for H + 1.7 Å

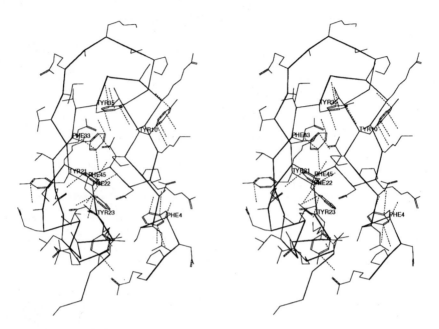

Figure 11.2. Global view of BPTI showing CH/π interactions (stereo). Thick lines indicate the α carbon plot (hydrogens are omitted).

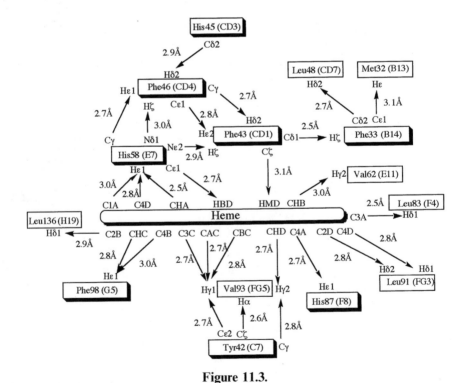

Figure 11.3.

for C) $\times 1.05 = 3.05$ Å] with the appropriate angle parameters are shown in Figure 11.3 for an α subunit of heme group of horse deoxyhemoglobin.[14]

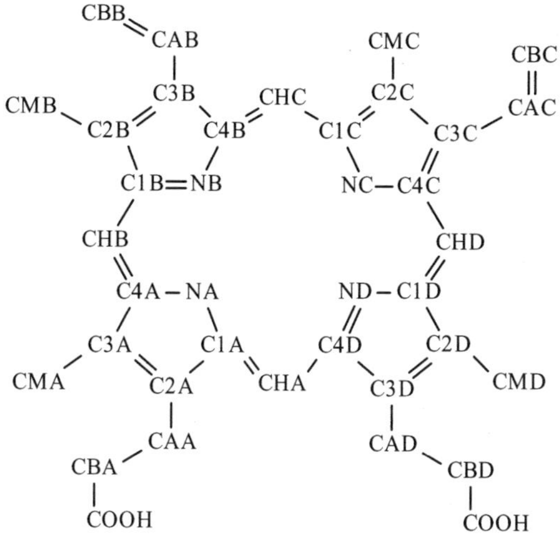

Heme

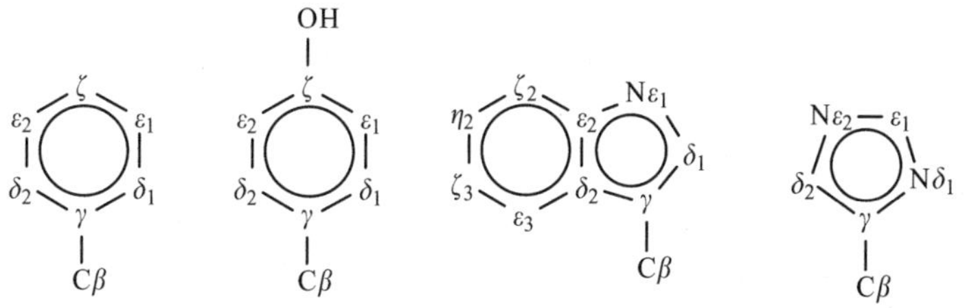

Numbering of atoms in side-chains of aromatic amino acids

Here we see many contacts shorter than the van der Waals distance. The CH/π bonds seem to play an important part in holding the heme group and stabilizing the protein structure. Histidine behaves as an aromatic residue. Sigel et al. reported[16] that the imidazole group plays a role similar to the phenyl or indole groups in intramolecular ligand/ligand interactions of ternary metal complexes. The tendency

Figure 11.3. CH/π interactions around the heme group in horse deoxyhemoglobin α subunit. The numbers refer to CH/C_{sp^2} distances (D_{atm}). The labels in parentheses are stereochemical notations by Kendrew et al.[15] Shaded rectangles represent aromatic residues. Greek letters indicate atoms relevant to the CH/π contact. Interatomic distances are shown for the closest ones among the atoms in the respective aromatic ring.

to form folded conformations was found to increase in the following order: imidazole < phenyl < indole. This implies that the size and basicity of the aromatic moiety is important in bringing about an effective association of the groups. The histidine side-chain is therefore anticipated to behave as an aromatic group, although less effectively than in those of tryptophane, tyrosine, or phenylalanine. Another interesting feature in Fig. 11.3 is that the two methyl groups in valine and leucine ($C\gamma1$ and $C\gamma2$ in Val93, and $C\delta1$ and $C\delta2$ in Leu91) are simultaneously involved in CH/π interactions with the porphyrin moiety. Interactions involving geminal methyl groups were also noted in subunit β.

11.4. INTERACTIONS INVOLVING CARBOHYDRATES

Quiocho et al.[17] studied the crystal structures of a number of periplasmic proteins, such as L-arabinose binding protein,[18] D-galactose binding protein (GBP),[19] D-maltose binding protein,[20] and complexes with their specific substrates. They found that, in every case, the carbohydrate ligands are sandwiched by aromatic side-chains of the proteins. Figure 11.4 gives the result of CHPI analysis for the GBP/glucose complex.[21]

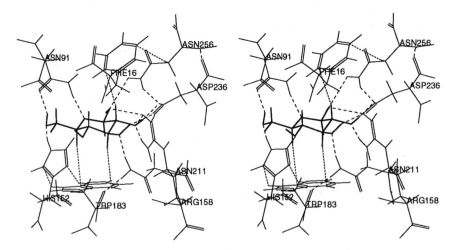

Figure 11.4. CH/π interactions in GBP/glucose complex (stereo view). Dotted and dashed lines indicate CH/C_{sp^2} contacts and hydrogen bonds, respectively.

Here we see many hydrogen bonds involving Asn91, Asp154 (not shown), Asn256, Asp236, Asn211, His152, and Arg158. In addition to these, the sugar is sandwiched by two aromatic residues. This was reported as a stacking of "hydrophobic" patches with C^3H, C^5H, and C^6H of glucose by Trp183 and C^2H and C^4H by Phe16.[22] A CHPI search demonstrated that $C\varepsilon3$ and $C\varepsilon2$ of Trp183 and $C\delta2$ of Phe16

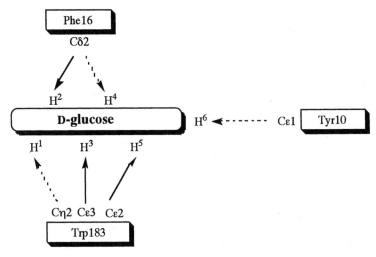

Figure 11.5. CH/C_{sp^2} contacts in a GBP/glucose complex.

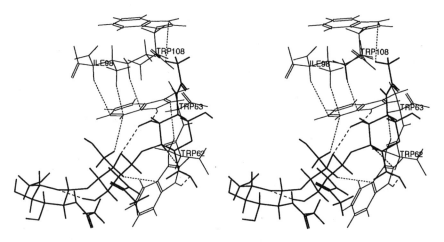

Figure 11.6. Interactions around the ligand in the lysozyme/(GlcNAc)$_3$ complex (stereo view). Dotted and dashed lines indicate CH/π contacts and hydrogen bonds, respectively.

are in CH/π contact with C^3H(2.8 Å), C^5H(3.0 Å), and C^2H(2.9Å) of the ligand, respectively.

Interactions Tyr10($C\varepsilon1$)/HC^6(3.2 Å), Phe16($C\delta2$)/HC^4(3.2 Å), and Trp183($C\eta2$)/HC^1(3.3 Å) were recorded when a longer cut-off value (D_{max}) of 3.4 Å was used in the survey (Fig. 11.5, denoted by dashed arrows). It seems that the axially oriented methine C–H bonds in glucose participated simultaneously in interactions with the sp^2 carbons of the aromatic rings.

Figures 11.6 and 11.7 show interactions found in a hen egg-white lysozyme complex with its specific substrate tri-*N*-acetylchitotriose.[23] A network involving a number of CH/π interactions is shown around the ligand, along with the hydrogen bonding present (Fig. 11.6).

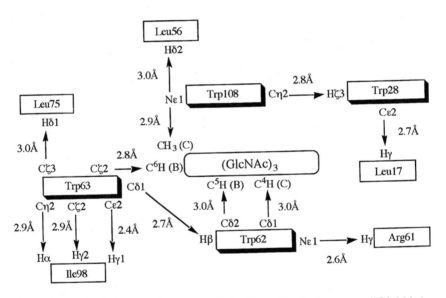

Figure 11.7. CH/π interactions around the ligand in the lysozyme/(GlcNAc)$_3$ complex.

tri-*N*-acetylchitotriose

The methyl group in the terminal acetamide in ring C is close to Trp108. Nε1 in Trp108 is in CH/π contact with a hydrogen in the acetyl group in ring C of the ligand. Cδ1 and Cδ2 in Trp62 are close to C^4H (ring C) and C^5H (ring B), respectively. Cζ2 of Trp63 is close to CH_2OH (ring B). Examination by CHPI demonstrated that interactions of the ligand with Trp108, Trp62, and Trp63 are assisted by a CH/π network involving Leu56, Trp28, Leu17, Arg61, Leu75, and Ile98 (Fig. 11.7). A study with enzymes prepared by point mutation has shown that the residue in position 63 (Trp in avian and Tyr in human lysozyme) must be an aromatic one.[24] Muraki et al. compared the enzymatic activities and crystal structures of Y63F (Tyr63 was converted to Phe. Y: tyrosine, F: phenylalanine), Y63W (W: tryptophane), Y63L (L: leucine), and Y63A (A: alanine) with human lysozyme. Stability and enzymatic activity of Y63F and Y63W are comparable to those of the wild enzyme, whereas those of Y63L and Y63A are not.

Xylose isomerase catalyses the isomerization of D-xylose. D-Sorbitol is a specific inhibitor of this enzyme, since the molecule closely resembles an open-chain configuration involved in the transition state of the reaction. Blow et al. studied the crystal structures of complexes of xylose isomerase with specific substrates and inhibitors.[25] Figures 11.8 and 11.9 show part of the structure, as disclosed by a CHPI analysis for the protein/D-sorbitol complex.

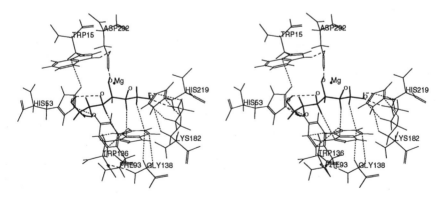

Figure 11.8. Interactions in xylose isomerase/D-sorbitol complex (stereo view). Dotted and dashed lines indicate CH/π contacts and hydrogen bonds, respectively.

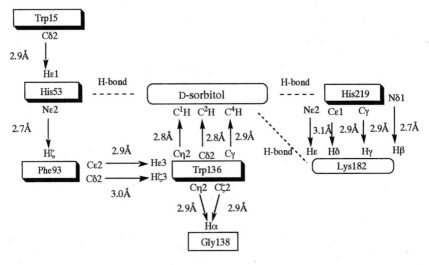

Figure 11.9. Interactions around the ligand in a xylose isomerase/D-sorbitol complex.

$$C^1H_2OH$$
$$|$$
$$H-C^2-OH$$
$$|$$
$$HO-C^3-H$$
$$|$$
$$H-C^4-OH$$
$$|$$
$$H-C^5-OH$$
$$|$$
$$C^6H_2OH$$

D-sorbitol

The atoms in the sorbitol molecule show a linear arrangement and the indolyl ring of Trp136 is found to be lined with a consecutive arrangement of the CH groups of the inhibitor. The temperature factors of the inhibitor were reported to be low, indicating that the interaction involved is strong. Short CH/π contacts are seen for Trp136 with the ligand ($C^4H/C\gamma$, $C^2H/C\delta2$, and $C^1H/C\eta2$). Trp136 is supported by interactions with Gly138 and Phe93. This in turn is assisted by His53 and then by Trp15. These interactions stabilize the structure of the complex, in cooperation with a number of hydrogen bonds and salt bridges around the ligand and the magnesium ion. Only one CH/π interaction, on the other hand, has been found (Trp15 $C\zeta2/HC^3$, 2.6 Å by a CHPI survey) in the xylose isomerase complexed with a cyclic sugar substrate.[26]

11.5. INTERACTIONS INVOLVING LYSINE AND ARGININE

The G proteins constitute a family of guanine-nucleotide binding proteins that serve as transducers of intracellular signaling pathways.[27] Ras p21 is an oncogene product involved in the growth-promoting signal-transduction system, an important member of the G proteins. Figure 11.10 shows a part of the CHPI analysis on the crystal structure of ras p21, complexed with GppNp, a GTP analog.[28]

GppNp

Asp119 and Asn116 are known to be vital for high G specificity due to hydrogen bonding with the guanine base. Besides these, there are two lysines, Lys117 and Lys147, sandwiching the guanine aromatic ring (Fig. 11.11). Lys117 is in CH/π contact at two points with guanine ($H\varepsilon/N^9$, $H\gamma/C^5$). Lys147 also interacts with guanine ($H\beta/N^1$). In the Ramachandran diagram, Lys117 was reported to locate itself at a region where few residues other than glycine are found, indicating that this is in an unusual conformation. The guanine aromatic ring is in contact with Phe28 ($N^3/H\zeta$, $C^5/H\varepsilon$); the latter is in turn assisted by two further CH/π interactions with Lys147 ($C\gamma/H\alpha$, $C\delta2/H\gamma$). Low-temperature factors were recorded for these residues in the crystallographic determination. These interactions certainly contribute to the G specificity of p21.

The Src homology-2 (SH2) domains are modules of approximately 100 amino acid residues. A number of cytoplasmic proteins in a signaling pathway contains the SH2 sequence.[29] They bind phosphotyrosine (pTyr)-containing peptides with high affinity, recognizing pTyr and the adjacent polypeptide sequences. Kuriyan and co-workers studied the crystal structures of the SH2 domain of a v-Src protein (a viral oncogene product with tyrosine kinase

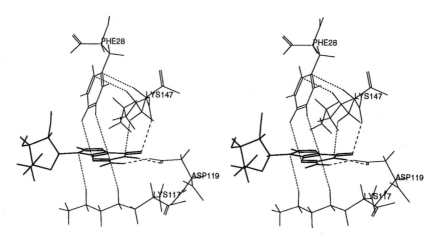

Figure 11.10. Stereo view of the guanine-binding region of ras p21/GppNp complex. Dotted and dashed lines indicate CH/C_{sp^2} contacts and hydrogen bonds, respectively.

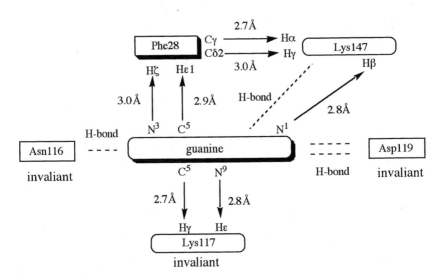

Figure 11.11. Guanine-binding region of ras p21/GppNp complex.

activity) complexed with tyrosine phosphorylated peptides.[30] Lys203 in a Src SH2 protein was reported to form a "hydrophobic" interaction with pTyr in a phosphorylated peptide, pTyr-Val-Pro-Met-Leu.

Examination by CHPI (Figs. 11.12 and 11.13) revealed that CHs (Hβ, Hε) in Lys203 are in contact with the sp^2 carbons of pTyr, while the NHs in the ε-amino group are found to be remote (>3.3 Å) even from the nearest C_{sp^2}. The guanidinium group in Arg155 side-chain

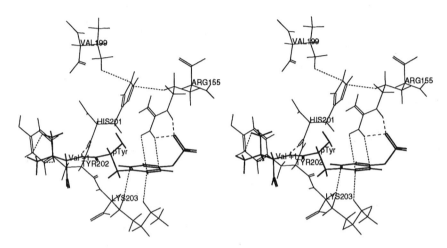

Figure 11.12. Stereo view of the tyrosine-binding region of Src SH2 domain/ phosphotyrosyl peptide [pTyr-Val-Pro-Met-Leu, peptide A] complexes. Dotted and dashed lines indicate CH/π contacts and hydrogen bonds, respectively.

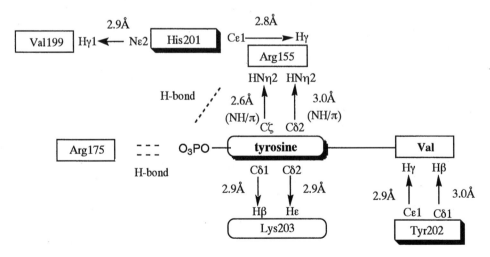

Figure 11.13. CH/π interactions in Src SH2 domain/phosphotyrosyl peptide (pTyr-Val-Pro-Met-Leu) complex.

interacts with the aromatic ring of pTyr (two NH/π interactions).[31] His201 is in CH/π contact with Val199 and Arg155. The valine side-chain next to pTyr is in CH/π contact with Tyr202.

Interaction between lysine and aromatic groups is found in many other proteins. This is, hereafter, referred to as the lysine CH/π interaction. Figures 11.14–11.16 give several examples of the lysine CH/π interaction from lysozyme and xylose isomerase.

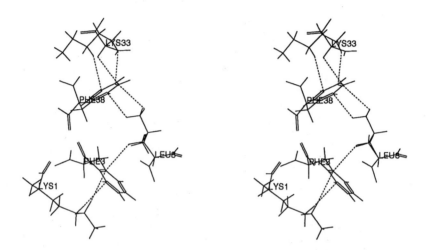

Figure 11.14. Lysine CH/π interactions in lysozyme (stereo view).

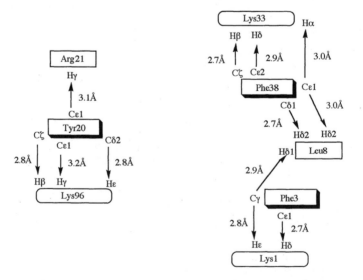

Figure 11.15. Lysine CH/π interactions in lysozyme.

A similar type of interaction was found in the complex of ε-amino-caproic acid (a lysine analog) with human plasminogen kringle.[32] The ligand is supported by Trp60 and Trp70, with four CH/π contacts each (Fig. 11.17).

Arginine seems to behave like lysine, with the use of three methyl-enes and the terminal guanidyl group. Figure 11.18 presents an

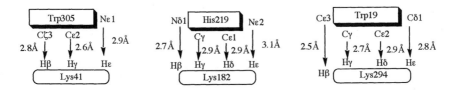

Figure 11.16. Lysine CH/π interactions in xylose isomerase.

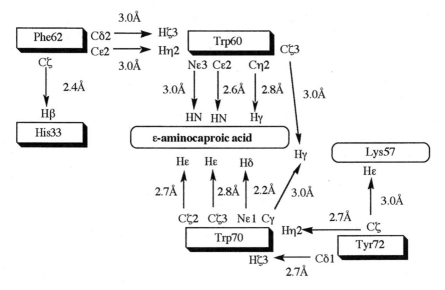

Figure 11.17. CH/π network around the ligand in human plasminogen kringle/ε-aminocaproic acid complex.

ε-aminocaproic acid

example for a sequence of residues in hGHbp:[33] Lys179/Trp186/Arg211/Phe225/Arg213/Tyr222/Lys215. The methylene hydrogens of lysines and arginines are involved in a linear arrangement.[34]

In most cases, multiple pairs of atoms are involved in the interaction. Four and three methylenes are present, respectively, in lysine and arginine side-chains between Cα and the terminal amino or guanidinium group. Thus, lysine and arginine have many chances to stabilize the structure of proteins and protein/ligand complexes by the CH/π interactions present in their structures (Fig. 11.19).

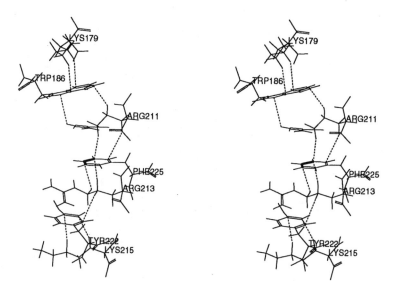

Figure 11.18. Lysine–arginine CH/π interactions in human growth hormone binding protein (stereo view).

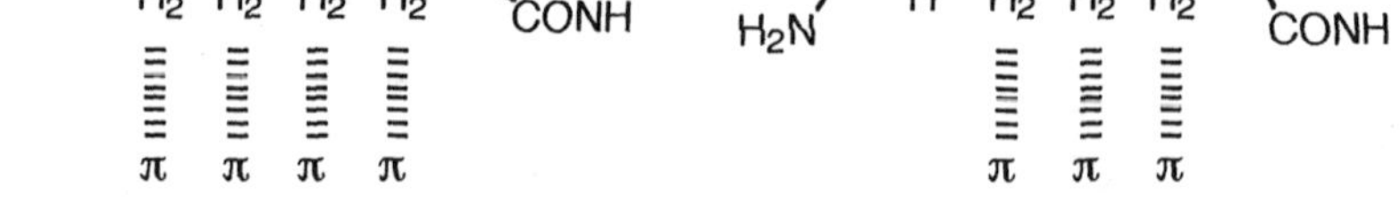

Figure 11.19. Lysine (arginine) CH/π interaction.

11.6. COMMENTS ON THE SO-CALLED CATION/π INTERACTION

An immunoglobulin fragment Fab McPC603 has a specific affinity to phosphorylcholine. For a rather long time, this specificity has been attributed to Coulombic interactions between positive (ammonium) versus negative (phosphoryl) charges in the complex.[35]

Figure 11.20 shows the result of a CHPI analysis around phosphorylcholine in McPC603.[36] Cγ, Nε1, Cε3, and Cζ2 in Trp107H (H; heavy chain), and Cζ in Tyr100L (L: light chain) are seen to be in contact with the hydrogens of CH_3N^+. Cζ2 in Trp107H and Cε1 in Tyr100L are close to a methylene hydrogen in the ligand.

$$H_3C - \overset{\overset{\textstyle CH_3}{|}}{\underset{\underset{\textstyle CH_3}{|}}{N^+}} - CH_2 - CH_2 - O - \overset{\overset{\textstyle O}{\|}}{\underset{\underset{\textstyle OH}{|}}{P}} - OH$$

phosphorylcholine

Dougherty and Stauffer discussed the problem on the basis of an attractive interaction between an ammonium cation and a π-electron

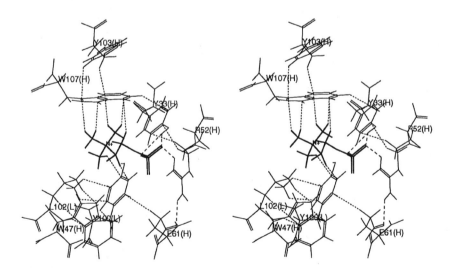

Figure 11.20. Interactions involving the ligand in McPC603/phosphorylcholine complex (stereo view). Dotted and dashed lines indicate CH/π contacts and hydrogen bonds, respectively.

system (cation/π interaction).[37] According to their argument, an electrostatic interaction of Me_3N^+ in the ligand with the negatively charged surface of the aromatic rings in the protein plays a central role. Coulombic or polar interactions of a similar nature may be important. This phenomenon, however, is better accommodated in the context of the CH/π interaction. To be effective, CH hydrogens need not necessarily be polarized.[38] The kinetics of the binding of acetylcholine (ACh) analogs such as 3,3-dimethylbutyl acetate, 4-t-butylthio-2-butanone, or 3,3-dimethylbutanol were studied. These neutral compounds bind as effectively as ACh, to the same subsite of the enzyme, acetylcholine esterase (AChE).[39] This demonstrates that the positive charge makes little contribution if any to the binding. Comparisons of quaternary compounds (Me_3N^+–R and Me_3C–R) as ligands with their corresponding lower analogs (Me_2N^+H–R and Me_2CH–R) showed the former to be more effective than the latter with regard to its binding capabilities to AChE. From this, the binding force involved here can be understood in terms of the CH/π interaction; the number and probability of the CHs participating in the interaction will have an appreciable effect in stabilizing the structure of the complex.

11.7. AROMATIC/AROMATIC INTERACTION

It has long been known that many aromatic groups interact favorably with each other in protein structures. In 1985, Burley and Petsko,[40] and Singh and Thornton,[41] independently studied the problem by examining a large number of protein coordinates deposited in PDB. They concluded that an approximate edge-to-face (T- or L-shape) arrangement of the aromatic rings is dominant. Explanations based primarily on electrostatic interaction (positive CH versus negative π-cloud) were proposed.[42] The nature of this favorable edge-to-face aromatic/aromatic interaction has, however, remained as an open question. Nevertheless, these favorable aryl/aryl interactions appear to be used in maintaining the stable and compact structures of proteins, particularly those of the lower molecular weight proteins.

Carp parvalbumin (109 residues), for instance, has neither cysteines nor metal atoms to stabilize its structure but is found stable with a core consisting of 10 phenylalanine residues.[43] To determine whether this is a possible case of CH/π interaction, a CHPI analysis was carried out for carp parvalbumin. Figures 11.21 and 11.22 give the results.[44]

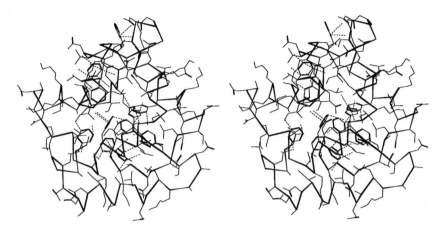

Figure 11.21. CH/π interactions in carp parvalbumin (stereo view; thick lines are the α carbon plot, hydrogens are omitted). Dotted lines indicate CH/C_{sp^2} contacts.

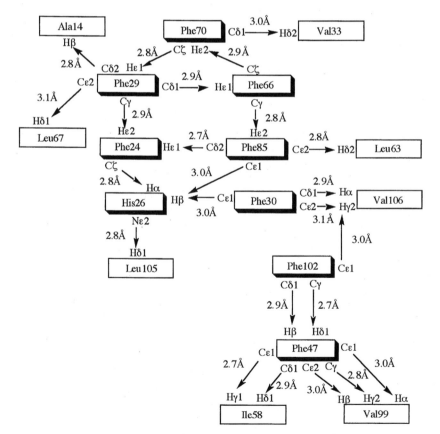

Figure 11.22. CH/C_{sp^2} contacts (D_{atm}) shown in carp parvalbumin.

Eight phenylalanines and a histidine have been found to participate in a large CH/π-network. There, an aromatic ring is often sandwiched by CH hydrogens of other aromatic residues. The contacts are shorter than the van der Waals distance, suggesting that the nature of the interaction is CH/π in type. CHs in an aromatic ring are more acidic than those in aliphatic groups and therefore are more prone to binding with π-bases. Sussman et al. reported that the active site of acetylcholine esterase lies at the bottom of a deep gorge lined with 14 aromatic residues.[45] In addition, the antigen-binding regions of certain immunoglobulin fragments were reported to have a close-packed cluster of aromatic side-chains.[46] It is likely that a specific substrate is guided by the aromatic residues lining the path to the gorge. It is tempting to speculate that an interaction of this kind may govern not only ligand binding, but also the folding process of macromolecules. This is reminiscent of the finding of Stoddart that the CH/π interaction acts as a driving force for the self-assembly of complex molecules such as pseudorotaxanes or catenanes.

11.8. OTHER POSSIBILITIES

The three-dimensional structures of D-glyceraldehyde 3-phosphate dehydrogenase from *Bacillus stearothermophilus*[47] and lobster[48] were determined. Preliminary search by CHPI revealed that the cofactor NAD in the former thermostable enzyme is supported by a number of CH/π interactions, whereas in the latter there are fewer CH/π contacts. It is possible that the heat stability of thermophilic proteins[49] finds its origin, at least partly, in the CH/π interaction.

Major histocompatibility complex (MHC) antigens are cell-surface glycoproteins that play an essential role in the cellular immune response. Class I MHC molecules are noncovalently associated dimers of two proteins: a heavy chain (α1, α2, and α3 domains of ca. 90 residues each) and a β2-microglobulin (ca. 100 residues). The α1 and α2 domains form a groove at the top portion of the protein; this groove is responsible for complex formation with various peptides.[50] The crystal structures were determined for a human class I MHC, HLA-A2 (HLA: human leukocyte antigen), and its complexes with peptides of viral origin.[51] CHPI analyses for complexes of HLA-A2 with five viral peptides have revealed that there are a number of CH/C_{sp^2} contacts around the ligand binding site and the subunit interfaces in the HLA protein.[52]

The Src homology-3 (SH3) domains are modules of approximately 60 residues found in a number of proteins in the intracellular signal-transduction system.[53] They bind with the highly nonpolar proline-rich domain(s) of target proteins in a specific manner. Structures of several SH3/ligand complexes were reported.[54] It is known that the overall topology is conserved regardless of variations in the SH3 and the proline-rich peptide sequences. All of the nonpolar contacts seem to be in line with our hypothesis that the CH/π interaction is playing an important role.

The crystal structure of a DNA-methyltransferase, M. Hhal, complexed with S-adenosyl-L-methionine was determined.[55] The conserved Phe18 and Pro80 are reported to form a hydrophobic platform of purine and ribose rings. A preliminary survey by CHPI has revealed a number of CH/C_{sp^2} contacts around the relevant groups.[56]

Uracil-DNA glycosylase (UDG) is a key enzyme in the DNA repair system. The crystal structure was studied for a complex of a virus UDG with uracil.[57] The uracil molecule was found to be stacked over the aromatic ring of an invariant phenylalanine (Phe101) and reported to be in van der Waals contacts with an invariant tyrosine (Tyr 90). The crystal structure of a 6-aminouracil complex of a human UDG was also studied.[58] A tyrosine residue (Tyr147) in the enzyme is found to be fixed with conserved residues, Val160 and Pro167. The side-chain of the aromatic ring of Tyr147 was reported to be in a T-shaped geometry with respect to the uracil plane and perhaps plays a role in discriminating its specific substrate.

To summarize this chapter, a huge body of evidence suggesting the importance of CH/π interactions in protein structures is available from PDB.

REFERENCES

1. M. Nishio, *Kagaku no Ryoiki*, **31**, 998 (1977); *ibid.*, **33**, 422 (1979); 29th Symposium on Protein Structures, Osaka (1978), Abstract, p. 161; S. Zushi, Y. Kodama, K. Nishihata, K. Umemura, M. Nishio, J. Uzawa, and M. Hirota, *Bull. Chem. Soc. Jpn.*, **53**, 3631 (1980).

2. M. Yun, C. Park, S. Kim, D. Nam, S. C. Kim, and D. H. Kim, *J. Am. Chem. Soc.*, **114**, 2281 (1992).

3. N. Morisaki, H. Funabashi, R. Shimazawa, J. Furukawa, A. Kawaguchi, S. Okuda, and S. Iwasaki, *Eur. J. Biochem.*, **211**, 111 (1993).

4. T. Tanaka, M. Hirama, K. Fujita, S. Imajo, and M. Ishiguro, *J. Chem. Soc., Chem. Commun.*, 1205 (1993).

5. K. H. Kim, B. M. Kwon, A. G. Myers, and D. C. Rees, *Science*, **262**, 1042 (1993).

6. K. Takahashi, T. Tanaka, T. Suzuki, and M. Hirama, *Tetrahedron*, **50**, 1327 (1994). Review: M. Hirama, *J. Synth. Org. Chem. Jpn.*, **52**, 980 (1994).

7. K. Jitsukawa, K. Iwai, H. Masuda, H. Ogoshi, and H. Einaga, *Chem. Lett.*, 303 (1994)

8. C. A. McPhalen, M. G. Vincent, and J. N. Jansonius, *J. Mol. Biol.*, **225**, 495 (1992).

9. P. Chakrabarti and U. Samanta, *J. Mol. Biol.*, **251**, 9 (1995).

10. M. M. Flocco and S. L. Mowbray, *J. Mol. Biol.*, **235**, 709 (1994).

11. M. Nishio, Y. Umezawa, M. Hirota, and Y. Takeuchi, *Tetrahedron*, **51**, 8665 (1995).

12. 5PTI (1.0 Å for X-ray, 1.8 Å for neutron data): A. Wlodawer, J. Walter, R. Huber, and L. Sjolin, *J. Mol. Biol.*, **180**, 301 (1984).

13. The optimization was carried out by molecular mechanics energy minimization with the program KOPT. K. Kamiya and H. Umeyama, unpublished; A. Kajihara, H. Komooka, K. Kamiya, and H. Umeyama, *Protein Eng.*, **6**, 615 (1993). The force-field parameters of the AMBER version 3.0, revision A were used. See S. J. Weiner, P. A. Kollman, D. T. Nguyen, and D. A. Case, *J. Comput. Chem.*, **7**, 230 (1986).

14. 2DHB (2.8 Å): W. Bolton and M. F. Perutz, *Nature*, **228**, 551 (1970). The hydrogen positions calculated on the basis of low-resolution diffraction data are not enough accurate in discussing the interatomic distances precisely. Discussions regarding the H/π distances on the protein structures obtained by low-resolution studies remain qualitative.

15. J. C. Kendrew, H. C. Watson, B. E. Strandberg, R. E. Dickerson, D. C. Phillips, and V. C. Shore, *Nature*, **190**, 663 (1961).

16. R. Malini-Balakrishnan, K. H. Scheller, U. K. Haring, R. Tribolet, and H. Sigel, *Inorg. Chem.*, **24**, 2067 (1985); H. Sigel, R. Tribolet, and O. Yamauchi, *Comments Inorg. Chem.*, **9**, 305 (1990).

17. Review: F. A. Quiocho, *Ann. Rev. Biochem.*, **55**, 287 (1986).

18. F. A. Quiocho and N. K. Vyas, *Nature*, **310**, 381 (1984).

19. N. K. Vyas, M. N. Vyas, and F. A. Quiocho, *Nature*, **327**, 635 (1987).

20. J. C. Spurlino, G.-Y. Lu, and F. A. Quiocho, *J. Biol. Chem.*, **266**, 5202 (1991).

21. N. K. Vyas, M. N. Vyas, and F. A. Quiocho, *Science*, **242**, 1290 (1988).

22. 3GBP (2.5 Å): S. L. Mowbray, R. D. Smith, and L. B. Cole, *Receptor*, **1**, 41 (1990).

23. 1HEW (1.8 Å): C. Cheetham, P. J. Artymiuk, and D. C. Phillips, *J. Mol. Biol.*, **244**, 613 (1992); C. C. F. Blake, R. Cassels, C. M. Dobson, F. M. Poulsen, R. J. P. Williams, and K. S. Wilson, *ibid.*, **147**, 73 (1981).

24. M. Muraki, K. Harata, and Y. Jigami, *Biochemistry*, **31**, 9212 (1992).

25. 4XIA (2.3 Å): K. Henrick, C. A. Collyer, and D. M. Blow, *J. Mol. Biol.*, **208**, 129 (1989).

26. 4XIS (1.6 Å): M. Whiltlow, A. J. Howard, B. C. Finzel, T. L. Poulos, E. Winborne, and G. L. Gilliland, *Prot. Struct. Funct.*, **9**, 153 (1991).

27. Review: A. G. Gilman, *Ann. Rev. Biochem.*, **56**, 615 (1987).

28. 5P21 (1.35 Å): E. F. Pai, U. Krengel, G. A. Petsko, R. S. Goody, W. Kabusch, and A. Wittinghofer, *EMBO J.*, **9**, 2351 (1990).

29. Review: T. Pawson, *Nature*, **373**, 573 (1995).

30. 1SHA (1.5 Å), 1SHB (2.0 Å): G. Waksman, D. Kominos, S. C. Robertson, N. Pant, D. Baltimore, R. B. Birge, D. Cowburn, H. Hanafusa, B. J. Mayer, M. Overduin, M. D. Resh, C. B. Rios, L. Silverman, and J. Kuriyan, *Nature*, **358**, 646 (1992).

31. M. F. Perutz, *Phil. Trans. R. Soc. A*, **345**, 105 (1993).

32. 2PK4 (1.9 Å): A. M. Mulichak, A. Tulinsky, and K. G. Ravichandran, *Biochemistry*, **30**, 10576 (1991). Kringles are folding domains which occur repeatedly among proteins primarily involved in blood coagulation pathways.

33. 3HHR (2.6 Å): T. Clackson and J. A. Wells, *Science*, **267**, 383 (1995).

34. Y. Umezawa and M. Nishio, *Bioorg. Med. Chem.*, **6**, 493 (1998).

35. E. A. Padlan, D. R. Davies, S. Rudikoff, and M. Potter, *Immunochemistry*, **13**, 945 (1976).

36. 2MCP (3.1 Å): R. J. Poljak, *Nature*, **256**, 373 (1975); E. A. Padlan, G. H. Cohen, and D. R. Davies, *Ann Immunol. (Paris), Sect. C*, **136**, 271 (1985).

37. D. A. Dougherty and D. A. Stauffer, *Science*, **250**, 1558 (1990).

38. The hydrogens in Me_3N^+ are positively charged compared to those in normal aliphatic groups and therefore are more prone to CH/π interaction. Approximately an eightfold increase in the binding ability of Me_3N^+H from that of Me_3COH was reported for calixarene complexes.

39. F. B. Hasan, S. G. Cohen, and J. B. Cohen, *J. Biol. Chem.*, **255**, 3898 (1980); F. B. Hasan, J. L. Elkind, S. G. Cohen, and J. B. Cohen, *ibid.*, **256**, 7781 (1981); S. G. Cohen, D. L. Lieberman, F. B. Hasan, and J. B. Cohen, *ibid.*, **257**, 14087 (1982).

40. S. K. Burley and G. A. Petsko, *Science*, **229**, 23 (1985); S. K. Burley and G. A. Petsko, *J. Am. Chem. Soc.*, **108**, 7995 (1986).

41. J. Singh and J. M. Thornton, *FEBS Lett.*, **191**, 1 (1985); J. Singh and J. M. Thornton, *J. Mol. Biol.*, **211**, 595 (1990); C. A. Hunter, J. Singh, and J. M. Thornton, *ibid.*, **218**, 837 (1991).

42. T. Blundell, J. Singh, J. M. Thornton, S. K. Burley, and G. A. Petsko, *Science*, **234**, 1005 (1986); S. K. Burley and G. A. Petsko, *Adv. Protein Chem.*, **39**, 125 (1988).

43. 5CPV (1.6 Å): A. L. Swain, R. H. Kretsinger, and E. L. Amma, *J. Biol. Chem.*, **264**, 16620 (1989).

44. M. Nishio, Y. Umezawa, M. Hirota, and Y. Takeuchi, *Tetrahedron*, **51**, 8665 (1995).

45. 1ACE (2.8 Å): J. L. Sussman, M. Harel, F. Frolow, C. Oefner, A. Goldman, L. Toker, and I. Silman, *Science*, **253**, 872 (1991).

46. A. B. Edmundson, K. R. Ely, R. L. Girling, E. E. Abola, M. Schiffer, F. A. Westholm, M. D. Fausch, and H. F. Deutsch, *Biochemistry*, **13**, 3816 (1974); A. B. Edmundson, K. R. Ely, E. E. Abola, M. Schiffer, and N. Panagiotopoulos, *ibid.*, **14**, 3953 (1975); L. M. Amzel, R. J. Poljak, F. Saul, J. M. Varga, and F. F. Richards, *Proc. Natl. Acad. Sci. USA*, **71**, 1427 (1974); R. J. Poljak, L. M. Amzel, B. L. Chen, R. P. Phizackerley, and F. Saul, *ibid.*, **71**, 3440 (1974); J. Novotony and E. Haber, *ibid.*, **82**, 4592 (1985); R. Mizutani, K. Miura, T. Nakayama, I. Simada, Y. Arata, and Y. Satow, *J. Mol. Biol.*, **254**, 208 (1995).

47. T. Skarzynski and A. J. Wonacott, *J. Mol. Biol.*, **203**, 1097 (1988).

48. D. Moras, K. W. Olsen, M. N. Sabesan, M. Buehner, G. C. Ford, and M. G. Rossmann, *J. Biol. Chem.*, **250**, 9137 (1975).

49. M. F. Perutz and H. Raidt, *Nature*, **255**, 256 (1975); G. Biesecker, J. I. Harris, J. C. Thierry, J. E. Walker, and A. J. Wonacott, *ibid.*, **266**, 328 (1977); M. F. Perutz, *Science*, **201**, 1187 (1978).

50. P. J. Bjorkman and P. Perham, *Ann. Rev. Biochem.*, **59**, 253 (1990).

51. D. R. Madden, D. N. Garboczi, and D. C. Wiley, *Cell*, **75**, 693 (1993); L. J. Stern and D. C. Wiley, *Structure*, **2**, 245 (1994).

52. Y. Umezawa and M. Nishio, unpublished results.

53. Review: G. B. Cohen, R. Ren, and D. Baltimore, *Cell*, **80**, 237 (1995); T. Pawson, *Nature*, **373**, 573 (1995).

54. H. Yu, J. K. Chen, S. Feng, D. C. Dalgarno, A. W. Brauer, and S. L. Schreiber, *Cell*, **76**, 933 (1994); S. Feng, J. K. Chen, H. Yu, J. A. Simon, and S. L. Schreiber, *Science*, **266**, 1241 (1994); D. Kohda, H. Terasawa, S. Ichikawa, K. Ogura, H. Hatanaka, V. Mandiyan, A. Ullrich, J. Schlessinger, and F. Inagaki, *Structure*, **2**, 1029 (1994); A. Musacchio, M. Saraste, and M. Wilmanns, *Nat. Struct. Biol.*, **1**, 546 (1994); X. Wu, B. Knudsen, S. M. Feller, J. Zheng, A. Sali, D. Cowburn, H. Hanafusa, and J. Kuriyan, *Structure*, **3**, 215 (1995).

55. X. Cheng, S. Kumar, J. Posfai, J. W. Pflugrath, and R. J. Roberts, *Cell*, **74**, 299 (1993).

56. Y. Umezawa and M. Nishio, unpublished results.

57. R. Savva, K. McAuley-Hecht, T. Brown, and L. Pearl, *Nature*, **373**, 487 (1995).

58. C. D. Mol, A. S. Arval, G. Slupphaug, B. Cavli, I. Alseth, H. E. Krokan, and J. A. Tainer, *Cell*, **80**, 869 (1995).

CHAPTER 12

SUMMARY AND PROSPECTS

One characteristic feature of the CH/π interaction is its looseness. In other words, the CH/π interaction is favorable from the view point of entropy, as compared to the ordinary hydrogen bond or the OH/π interaction. This lends this weak attractive force, along with its multiple effect, a unique property to play unexpectedly potent roles in a variety of molecular interactions, often in cooperation with other secondary forces. The effect becomes more significant in compounds of high molecular weight bearing an abundance of CH and π groups. Moreover, unlike the conventional hydrogen bonds and the Coulombic force, the CH/π bond may play its role in protic media, such as in water, and by implication in living systems. A number of molecular phenomena attributed in the past to the van der Waals interaction (nonpolar, lipophilic interaction) or the so-called hydrophobic effect should now be reexamined in light of the new paradigm.

An enthalpy of a one-unit CH/π interaction is certainly the weakest among the interatomic interactions involving hydrogen, probably less than $1 \, \text{kcal mol}^{-1}$. This contrasts with other well-established weak hydrogen bonds such as the OH/π,[1] NH/π,[2] CH/O,[3] and CH/N[4] interactions. Perutz reviewed the role of aromatic rings as hydrogen bond acceptors of OH or NH.[5] The OH/π interaction mainly originates from charge-transfer interaction[6] or Coulombic force.[7] However, these forces seem to be relatively unimportant compared to the

CH/π bond when considering the dynamic aspects of molecular species, particularly in polar or protic solvents. The OH/π interaction (2–4 kcal mol^{-1}) is certainly stronger than the CH/π interaction, but it requires that the relevant groups be arranged in a manner unfavorable to dipole/quadrupole interaction. The OH/π interaction is entropically handicapped compared to the CH/π interaction. In Chapter 11 we note only a few OH/π bonds in proteins. An explanation may be that the OH group is prone to make hydrogen bonds only with strong bases. Another reason may be that the number of residues bearing side-chain groups with OH are only three (tyrosine, serine, and threonine) among the 20 amino acids, while the CHs are ubiquitous. The NH/π interaction (2–4 kcal mol^{-1}) is also stronger in enthalpy terms than is the CH/π interaction and this well-known type of interaction is often referred to as the amino aromatic interaction.[8]

As for the CH/n interaction (n: lone pair electrons), conformations of the smaller molecules such as propionaldehyde and its homologs (CH$_3$CHRCHO), haloalkanes (CH$_3$CHRCH$_2$X), and so on are found generally with the CH$_3$/Y (Y = O, N, or X) eclipsed or in a synclinal relationship. Formation of a five-membered chelate ring (CH/Y) may be suggested in these conformations (Fig. 12.1).

The NMR signals attributed to carbons gamma to a hetero atom (O, N, or X) in aliphatic alcohols, amines, or haloalkanes are shifted significantly to higher magnetic fields (^{13}C γ-effect). This may also be due, at least in part, to this type of interaction (Fig. 12.2).

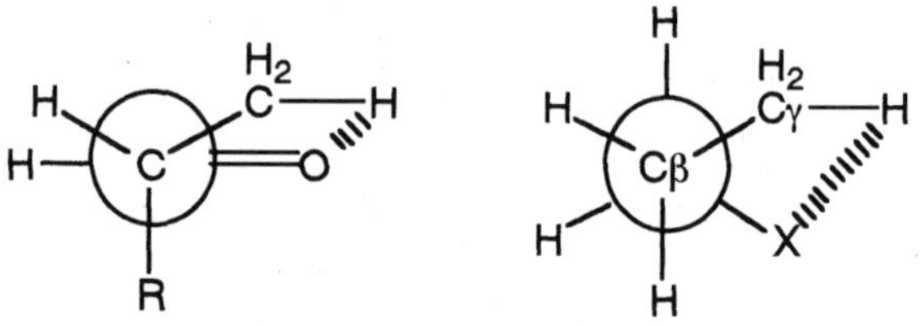

Figure 12.1. CH/n interactions in propionaldehyde and haloalkanes.

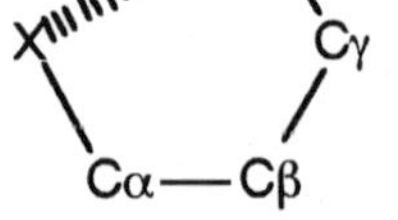

Figure 12.2. A possible mechanism for ^{13}C γ-effect.

CH/n interaction offers a wide-open area of future research. Derewenda et al. presented a discussion on the occurrence of CH/O hydrogen bonds in proteins.[9]

12.1. POSSIBILITIES IN BIOCHEMISTRY

In hemoglobin, we noticed that the geminal dimethyl groups in valine and leucine are involved simultaneously in the interaction with the heme group (Section 11.2). Although not particularly common, this type of interaction is also noted in other proteins. The interactions involving geminal (or vicinal) dimethyl groups versus those in aromatic π-systems illustrate the importance of combining a multiple number of interactions. This, along with other data[10] suggesting the remarkable role of the isopropyl (or, more generally branched alkyl) moiety in molecular recognitions, may explain why only amino acids bearing a branched aliphatic group (valine, leucine, and isoleucine) are present in proteins, whereas amino acids with a straight-chain alkyl group are not found in naturally occurring proteins (Fig. 12.3). Neither homoalanine (n-C_2H_5), norvaline (or aminobutyrate, n-C_3H_7), nor norleucine (n-C_4H_9) are present in nature.

One likely interpretation is that the amino acids with straight-chain groups are less capable of producing strong enough CH/π, CH/n, or van der Waals interactions compared to those present in branched ones and have thus dropped out under the "pressure" of natural selection.

Lysine and arginine are generally envisaged as responsible for the formation of salt-bridges or hydrogen bonds. However, it is apparent from considerations in Chapter 11 that the side-chains of these residues are good π-acceptors. The CHs in aromatic rings are also good π-acceptors. Amino acids may thus be categorized either as π-donating or π-accepting (Table 12.1).

From the chemical structure, the side-chains of lysine, arginine, threonine, asparagine, and glutamine can be regarded as π-accepting, along with those of alanine, valine, leucine, isoleucine, methionine, and proline. The aromatic groups of tryptophane, tyrosine, phenylalanine, and histidine are π-donating (decreasing in strength in this order) as well as π-accepting. In this context, α-CH and β-CH of every residue are π-acceptors.

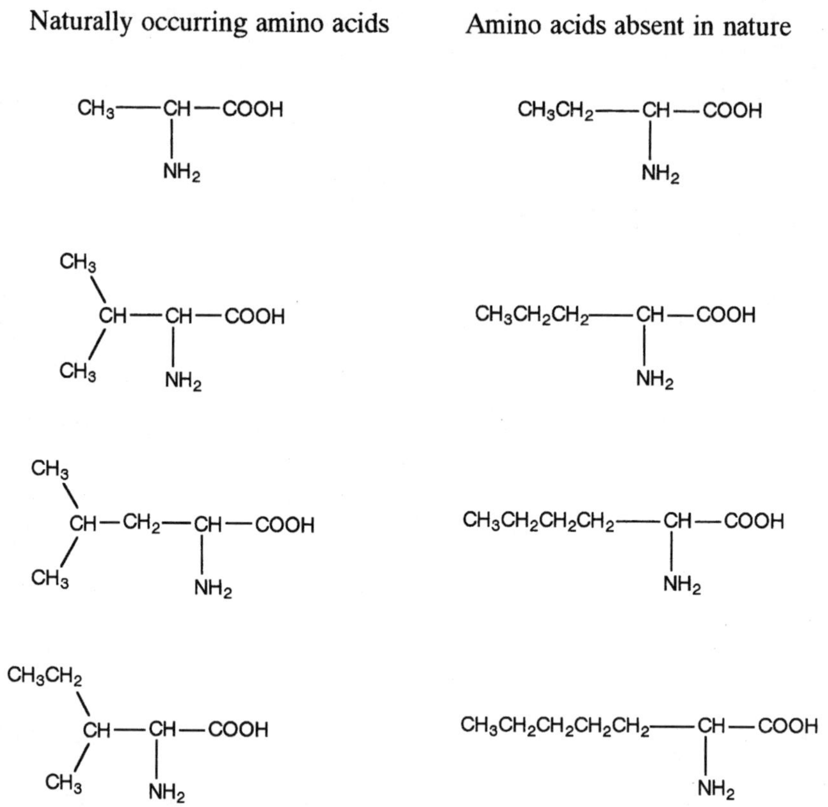

Naturally occurring amino acids   Amino acids absent in nature

CH₃——CH—COOH   CH₃CH₂——CH—COOH
        | |
       NH₂   NH₂

CH₃
   \
    CH—CH—COOH   CH₃CH₂CH₂——CH—COOH
   / | |
CH₃   NH₂   NH₂

CH₃
   \
    CH—CH₂—CH—COOH   CH₃CH₂CH₂CH₂——CH—COOH
   / | |
CH₃   NH₂   NH₂

CH₃CH₂
     \
      CH—CH—COOH   CH₃CH₂CH₂CH₂CH₂——CH—COOH
     / | |
   CH₃   NH₂   NH₂

Figure 12.3. Amino acids present in nature (left) and those not found in nature (right).

The CH/π interaction may be seen in interactions between proteins, proteins and carbohydrates, and proteins and nucleic acids. The CH/π attractive force may also play a role in interactions involving lipids. Biomembranes are composed primarily of a variety of lipids having saturated or unsaturated aliphatic groups. The dynamic interaction of phospholipid bilayers with proteins or organic compounds of biological importance may involve CH/π interactions. This concept may be useful in the engineering of macromolecules, including proteins, for a variety of purposes and in designing biologically active substances such as drugs and agrochemicals. In this regard it may be worthwhile to note that many compounds of practical medical use bear an aromatic moiety in the molecule.

TABLE 12.1. π-Donating and π-Accepting Residues

π-Donating	π-Accepting (H-donating)
Trp, Tyr, Phe, His	Ala, Val, Leu, Ile, Met, Pro
	Lys, Arg, Thr, Asn, Gln
	Trp, Tyr, Phe, His

12.2. POSSIBILITIES IN MATERIALS SCIENCE

Mori et al. found that the mesogenic properties of 5-alkoxy-2-(4-alkoxybenzoylamino)tropones are controlled by the length of the alkoxy group (RO) on the tropone ring.[11] Thus, ethoxy or butoxy ($R = C_2H_5$ or $n\text{-}C_4H_9$, $R' = OCH_3$) derivatives are mesogenic. However, an analog with $R = CH_3$ showed no mesogenic property. When R was $n\text{-}C_4H_9$, or the perfluorobenzoyl analog, or when an electron-withdrawing group was placed on the benzene ring ($R' = CN$ or NO_2), there was no longer any sign of a mesogenic property, while the derivatives with an electron-donating group ($R' = OC_2H_5$, $n\text{-}C_3H_7$, NMe_2) showed mesogenic properties.

5-alkoxy-2-(4-alkoxybenzoylamino)tropones

The structure of a derivative ($R = n\text{-}C_4H_9$, $R' = OCH_3$) in crystal is illustrated in Figure 12.4. Two molecules are arranged in a head-to-tail manner and CH/π interactions (CH/C 3.00 Å at the shortest) are found between the aliphatic part and the benzoyl group. The tropone rings are stacked (shortest CH/C distance, 3.52 Å).

The self-assembly system reported by Stoddart et al. (Sections 9.4 and 9.5) utilizes combinations of weak molecular forces such as the CH/π (aromatic as well as aliphatic) and CH/O interactions. Stacking of aromatic groups (electron-deficient rings versus electron-rich ones) is considered to be a driving force in the self-assembly process. Relevance of the "π/π stacking" is apparent, at least phenomenologically. This was argued in terms of electrostatic interaction. According to

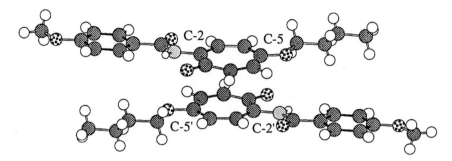

Figure 12.4. X-ray crystal structure of 5-butoxy-2-(4-ethoxybenzoylamino)-tropone.

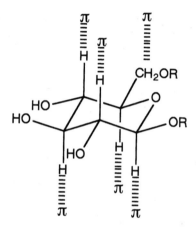

Figure 12.5. Speculated interactions between a synthetic dye (bearing many π-electrons) and a glucose moiety of cellulose.

Hunter and Sanders,[12] this type of interaction is favorable only when the aromatic groups adopt a parallel arrangement with their centers offset. We suggest that a charge-transfer interaction might also occur if the two π rings are displaced from each other at an appropriate distance and orientation, though the overlap of the relevant orbitals is not ideal with this geometry. This suggestion has by no means been verified and is open to question.

In 1966, Yoshida and Osawa pointed out the possibility that the staining of cotton by certain dyes (bearing many aromatic groups) may be a consequence of OH/π interaction.[13] They suggested that the OHs in the glucose moiety of the cellulose interact with the aromatic parts of the dye molecules. A more plausible explanation may be that this is due to CH/π interaction. The CHs in the glucose unit are oriented axially to the molecular plane and may interact favorably with the π-electrons of the dyes (Fig. 12.5).

Think of drawing a character (or a figure, etc.) with carbon black or graphite $[(\pi)_n$, n: numerous] on a sheet of paper. The paper is made of cellulose $(CHOH)_n$. Drawing a character with carbon onto paper is equivalent to forming an infinite number of $(CH/\pi)_n$ interactions. Try and erase this with any available material. Water does not work. Neither do organic solvents or brushes. Does it work with an eraser (made of rubber)? An eraser is principally made up of hydrocarbon polymers in which CHs exist in a more condensed manner than those present in a cellulose (Fig. 12.6).

Automobile tires are made of rubber, principally, mixed with fine powder of carbon black. Let us consider the kinds of interaction to which tires lend themselves. An intriguing speculation is that CH/π interaction is working, almost infinitely, between the CHs in rubber and the π groups in graphite, thus giving strong durability and elasticity (Fig. 12.7). How many more pairs of other types of materials can we imagine that give such a useful property?

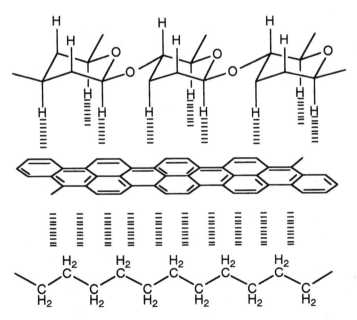

Figure 12.6. Competition of the CH/π interaction (suggested) between graphite and carbohydrate vs. graphite and hydrocarbon.

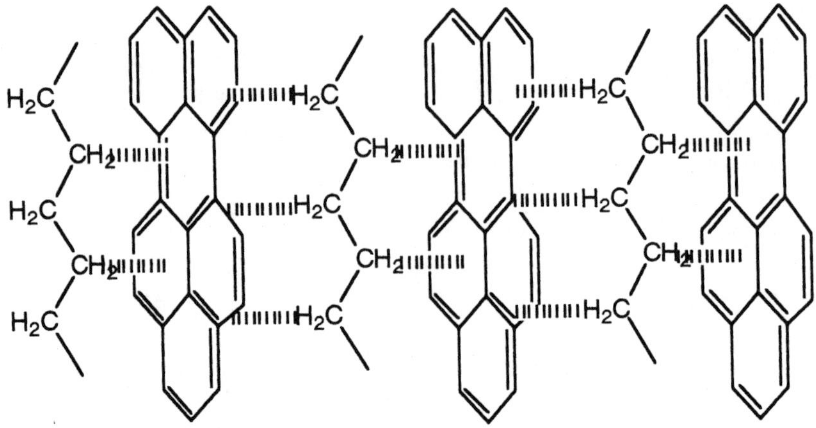

Figure 12.7. CH/π interaction (speculated) between graphite and hydrocarbon polymer.

CONCLUSION

Evidence, nature, and significance of the CH/π interaction have been discussed throughout the book, in light of our present knowledge. We conclude that the CH/π interaction is highly useful in understanding a wide range of molecular phenomena including life science and materials science.

REFERENCES

1. M. Oki and H. Iwamura, *Bull. Chem. Soc. Jpn.*, **32**, 1135 (1959); *ibid.*, **33**, 1600 (1960); H. M. Fales and W. C. Wilman, *J. Am. Chem. Soc.*, **85**, 784 (1963); A. T. McPhail and G. A. Sim, *Chem. Commun.*, 124 (1965); Z. Yoshida and E. Osawa, *Nippon Kagaku Zasshi*, **87**, 509 (1966); M. Ōki and H. Iwamura, *J. Am. Chem. Soc.*, **89**, 576 (1967); M. Oki, H. Iwamura, T. Onoda, and M. Iwamura, *Tetrahedron*, **24**, 1905 (1968); S. Ueji, *Bull. Chem. Soc. Jpn.*, **51**, 1799 (1978); S. Ueji, K. Nakatsu, H. Yoshioka, and K. Kinoshita, *Tetrahedron Lett.*, 1173 (1982); J. L. Knee, L. R. Khundkar, and A. H. Zewail, *J. Chem. Phys.*, **87**, 115 (1987).
2. W. Klempere, M. W. Cronyn, A. M. Maki, and G. C. Pimentel, *J. Am. Chem. Soc.*, **76**, 5846 (1954); M. Oki and K. Mutai, *Tetrahedron*, **26**, 1181

(1970); K. Mutai, *Tetrahedron Lett.*, 1125 (1971); C. A. Deakyne and M. Meot-Ner (Mautner), *J. Am. Chem. Soc.*, **107**, 474 (1985); P. N. Jagg, P. F. Kelly, H. S. Rzepa, D. J. Williams, J. D. Woollins, and W. Wylie, *J. Chem. Soc., Chem. Commun.*, 942 (1991); L. R. Hanton, C. A. Hunter, and D. H. Purvis, *ibid.*, 1134 (1992); S. W. Whiteheart, I. C. Griff, M. Brunner, D. O. Clary, T. Mayer, S. A. Buhrow, and J. E. Rothman, *Nature*, **362**, 353 (1993); D. A. Rodham, S. Suzuki, R. D. Suenram, F. J. Lovas, S. Dasgupta, W. A. Goddard, III and G. A. Blake, *ibid.*, **362**, 735 (1993); M. M. Flocco and S. L. Mowbray, *J. Mol. Biol.*, **235**, 709 (1994); P. K. Bakshi, A. Linden, B. R. Vincent, S. P. Roe, D. Adhikesavalu, T. S. Cameron, and O. Knop, *Can. J. Chem.*, **72**, 1273 (1994); K. S. Kim, J. Y. Lee, S. J. Lee, T. K. Ha, and D. H. Kim, *J. Am. Chem. Soc.*, **116**, 7399 (1994); J. Y. Lee, S. J. Lee, H. S. Choi, S. J. Cho, K. S. Kim, and T. K. Ha, *Chem. Phys. Lett.*, **232**, 67 (1995).

3. G. J. Korinek and W. G. Schneider, *Can. J. Chem.*, **35**, 1157 (1957); J. C. B. Brand, G. Eglinton, and J. F. Morman, *J. Chem. Soc.*, 2526 (1960); D. J. Sutor, *Nature*, **195**, 68 (1962); *J. Chem. Soc.*, 1105 (1963); A. Allerhand and P. von R. Schleyer, *J. Am. Chem. Soc.*, **85**, 1715 (1963); P. Kollman, J. McKervey, A. Johansson, and S. Rothenberg, *ibid.*, **97**, 955 (1975); H. Umeyama and K. Morokuma, *ibid.*, **99**, 1316 (1977); K. Morokuma, *Acc. Chem. Res.*, **10**, 294 (1977); S. Zushi, Y. Kodama, K. Nishihata, K. Umemura, M. Nishio, J. Uzawa, and M. Hirota, *Bull. Chem. Soc. Jpn.*, **53**, 3631 (1980); R. Taylor and O. Kennard, *J. Am. Chem. Soc.*, **104**, 5063 (1982); Y. Tamura, G. Yamamoto, and M. Ōki, *Bull. Chem. Soc. Jpn.*, **60**, 1781, *ibid.*, **60**, 3789 (1987); G. R. Desiraju, *J. Chem. Soc., Chem. Commun.*, 179 (1989); *ibid.*, 454 (1990); G. R. Desiraju and C. V. K. M. Sharma, *ibid.*, 1239 (1991); G. R. Desiraju, *Acc. Chem. Res.*, **24**, 290 (1991); *ibid.*, **29**, 441 (1996); T. Steiner and W. Saenger, *J. Am. Chem. Soc.*, **114**, 10146 (1992); *ibid.*, **115**, 4540 (1993).

4. M. Meot-Ner (Mautner) and C. A. Deakyne, *J. Am. Chem. Soc.*, **107**, 469 (1985); P. N. Jagg, P. F. Kelly, H. S. Rzepa, D. J. Williams, J. D. Woolins, and W. Wylie, *J. Chem. Soc., Chem. Commun.*, 942 (1991); C. Avendano, M. Espada, B. Ocana, S. Garcia-Granda, M. del R. Diaz, B. Tejerina, F. Gromez-Beltran, A. Martinez, and J. Elguero, *J. Chem. Soc., Perkin 2*, 1547 (1993).

5. M. F. Perutz, *Phil. Trans. R. Soc. A*, **345**, 105 (1993).

6. H. Iwamura, *Kagaku to Kogyo*, **17**, 617 (1963).

7. M. Levitt and M. F. Perutz, *J. Mol. Biol.*, **201**, 751 (1988).

8. A. Wlodawer, J. Walter, R. Huber, and L. Sjolin, *J. Mol. Biol.*, **180**, 301 (1984); M. F. Perutz, G. Fermi, D. J. Abraham, C. Poyart, and E. Bursaux, *J. Am. Chem. Soc.*, **108**, 1064 (1986); S. K. Burley and G. A. Petsko, *FEBS Lett.*, **203**, 138 (1986); T. M. Fong, M. A. Cascieri, H. Yu, A. Bansal, C. Swain and C. D. Strader, *Nature*, **362**, 350 (1993).

9. Z. S. Derewenda, L. Lee, and U. Derewenda, *J. Mol. Biol.*, **252**, 248 (1995).

10. The presence of a branched-alkyl group seems to be important for bringing about a significant effect in molecular recognitions. This is shown in a number of topics discussed in Chapters 2 and 5–11.

11. A. Mori, K. Hirayama, N. Kato, A. Takeshita, and S. Ujiie, *Chemistry Lett.*, 509 (1997).

12. C. A. Hunter and J. K. M. Sanders, *J. Am. Chem. Soc.*, **112**, 5525 (1990); C. A. Hunter, *Angew. Chem. Int. Ed.*, **32**, 1584 (1993); C. A. Hunter, *Chem. Soc. Rev.*, 101 (1994).

13. Z. Yoshida and E. Osawa, *Nippon Kagaku Zasshi*, **87**, 509 (1966).

INDEX

Protein folding, 198
Pseudorotaxane, 3, 152
Pyridine, 74
Pyridoxal-5-phosphate, 177
Pyrocalciferol, 35, 108
Pyrrole, 81

Ras p21 protein, 189, 190
Refractometric analysis, 5
Remote functionalization, 122
Ring size effect, 66, 84–85
Rotational strength, 32
Rotaxane, 154

Self assembly, 2, 154, 198, 207
Signal-transduction system, 199
Soft acids and soft bases, interaction of, 2, 68
Solvent polarity, effect of, 68, 88, 95, 165, 167
Somatostatin, 101
Speleand, 143
Src homology-2 domain (SH2), 189, 191
Src homology-3 domain (SH3), 199
Stacking, 154, 163, 207
Stereoselective reduction, 120
Steric compression, 15, 16
Steric requirement of intramolecular interaction
  CH/n interaction, 204
  CH/π interaction, 6, 107
  XH/π interaction, 84
Substituent effect
  as probes to assess the nature of CH/π interaction, 61–65
  atropisomerism, 20, 95, 168
  complex formation, 40

conformational equilibria, 17, 20, 29, 30, 37, 95
mesogenic property, 207
NOE enhancement, 29
rotational barrier, 95
stereoselectivity, 6, 119, 127, 128
Supramolecular chemistry, 131

Thermochemical measurement, 11
Toluene, 7, 14, 64, 81
Toluenes (substituted), 23
Tribromomethane, 74, 82
Trichloromethane (CHCl$_3$ and CDCl$_3$), 14, 21, 41, 56, 74, 81
Tricyclic orthoamide, 79
Trimethoxymethane, 82
Triptycenes, 20, 62
Tryptophane, 161, 184, 187
T-shaped geometry, 48, 53, 55, 94, 196, 199. *See also* Aromatic CH/π interaction
Tyrosine, 184, 187,199

Uracil-DNA glycosylase, 199

Valine, 184, 199, 205
van der Waals distance, 12, 33, 35, 49, 51, 59, 75, 181, 198
van der Waals force, 1, 49, 203
Vapor pressure measurement, 40
ω-Vinylalkan-1-ol, 84
v-Src protein, 189

XH/π interaction, 2
X-ray crystal structure, *see* Crystal structure
X-ray diffraction, 12, 33–36, 38, 75–77, 85–87
Xylene, 14, 64, 81
Xylose isomerase, 187, 188, 192